The Art and Craft of Mathematical Problem Solving

Part II

Professor Paul Zeitz

THE TEACHING COMPANY ®

PUBLISHED BY:

THE TEACHING COMPANY
4840 Westfields Boulevard, Suite 500
Chantilly, Virginia 20151-2299
1-800-TEACH-12
Fax—703-378-3819
www.teach12.com

Printed in the United States of America

ISBN 1-59803-622-X

Paul Zeitz, Ph.D.

Professor of Mathematics, University of San Francisco

Paul Zeitz is Professor of Mathematics at the University of San Francisco. He majored in History at Harvard University and received a Ph.D. in Mathematics from the University of California, Berkeley, in 1992, specializing in Ergodic Theory. Between college and graduate school, he taught high school mathematics in San Francisco and Colorado Springs.

One of Professor Zeitz's greatest interests is mathematical problem solving. He won the USA Mathematical Olympiad (USAMO) and was a member of the first American team to participate in the International Mathematical Olympiad (IMO) in 1974. Since 1985, he has composed and edited problems for several national math contests, including the USAMO. He has helped train several American IMO teams, most notably the 1994 "Dream Team," which was the first—and heretofore only—team in history to achieve a perfect score. This work, and his experiences teaching at USF, led him to write *The Art and Craft of Problem Solving* (Wiley, 1999; 2nd ed., 2007).

Professor Zeitz has also been active in events for high school students. He founded the San Francisco Bay Area Math Meet in 1994; cofounded the Bay Area Mathematical Olympiad in 1999; and currently is the director of the San Francisco Math Circle, a program that targets middle and high school students from underrepresented populations.

Professor Zeitz was honored in March 2002 with the Award for Distinguished College or University Teaching of Mathematics from the Northern California Section of the Mathematical Association of America (MAA), and in January 2003, he received the MAA's national teaching award, the Deborah and Franklin Tepper Haimo Award.

Table of Contents

The Art and Craft of Mathematical Problem Solving Part II

The Art and Craft of Mathematical Problem Solving

Scope:

This is a course about mathematical problem solving. The phrase "problem solving" has become quite popular lately, so before we proceed, it is important that you understand how I define this term.

I contrast problems with exercises. The latter are mathematical questions that one knows how to answer immediately: for example, "What is $3 + 8$?" or "What is 38^{74}?" Both of these are simple arithmetic exercises, although the second one is rather difficult, and the chance of getting the correct answer is nil. Nevertheless, there is no question about how to proceed.

In contrast, a problem is a question that one does not know, at the outset, how to approach. This is what makes mathematical problem solving so important, and not just for mathematicians. Arguably, all pure mathematical research is just problem solving, at a rather high level. But the problem-solving mind-set is important for all who take learning seriously, especially lifelong learners. Much of the current craze in brain strengthening focuses merely on exercises. These are not without merit—indeed, mental exercise is essential for everyone—but they miss out on a crucial dimension of intellectual life. Our brains are not just for doing crosswords or sudoku—they also can and should help us with intensive contemplation, open-ended experimentation, long wild goose chases, and moments of hard-earned triumph. That is what problem solving is all about.

An analogy that I frequently use compares an exerciser to a gym rat and a problem solver to a mountaineer. The latter's experience is riskier, messier, dirtier, less constrained, less certain, but much more fun. For those of you who prefer more civilized pursuits, consider 2 ways to learn Italian. One involves toiling over grammar exercises and translations of texts. The other method is to spend a few months, perhaps after a short bit of preparation, in a small town in Italy where no one else speaks English. Again, the latter approach is messier but fundamentally richer.

Becoming a good problem solver requires new skills (mathematical as well as psychological) and patient effort. My pedagogical philosophy is both experiential and analytic. In other words, you cannot learn problem solving without working hard at lots of problems. But I also want you to understand what you are doing at as high a level as possible. We will break down the process of solving a problem into investigation, strategy, tactics, and finer-grained tools, and we will often step back to discuss not just how we solved a problem but why our methods worked.

Problems, by definition, are hard to solve. Solving problems requires investigation, and successful investigations need strategies and tactics. Strategies are broad ideas, often not just mathematical, that facilitate investigation. Some strategies are psychological, others organizational, and others simply commonsense ideas that apply to problems in any field. Tactics are more narrowly focused, mostly mathematical ideas that help solve many problems that have been softened by good strategy. Additionally, there are very specialized techniques, called tricks by some, that I call tools.

This course is devoted to the systematic development of investigation methods, strategies, and tactics. Besides this "problemsolvingology," I will introduce you to mathematical folklore: classic problems as well as mathematical disciplines that play an important role in the problem-solving world. For example, no course on problem solving is complete without some discussion of graph theory, which is an important branch of math on its own but is also a very accessible laboratory for exploring problem-solving themes. Many of the lectures will include small amounts of new mathematics that we will build up and stitch together as the course progresses. The topics are largely drawn from discrete mathematics (graph theory, integer sequences, number theory, and combinatorics), because this branch of math does not require advanced skills such as calculus. That does not mean it is easy, but we will move slowly and develop new ideas carefully.

A small but important part of the course explores the culture of problem solving. I will draw on my experience as a competitor, coach, and problem writer for various regional, national, and international math contests, to make the little-known world of math Olympiads come to life. And I will discuss the recent educational reform movement (in which I am a key player) to bring Eastern European–inspired mathematical circles to the United States.

Problem solving is not a vertically organized discipline; it is not something that one learns in a linear fashion. Thus the overall organization of this course has a recursive, spiral nature. The first few lectures introduce the main ideas of strategy and tactics, which then are revisited and illuminated by different examples. We will often return to and refine previously introduced ideas. Overall, the topics get more complex toward the end of the course, but the underlying concepts do not really change. An analogy is a theme and variations musical piece, where the main theme is introduced with a slow, stately rhythm and later ends in complex avant-garde interpretations. By the end of the course, you should understand the main theme (the basic and powerful strategies and tactics of problem solving) quite well because you had to struggle with the complex interpretations (the advanced folklore problems that used the basic strategies in novel ways).

Problem solving is not just solving math problems. It is a mental discipline; successful investigations demand concentration and patient contemplation that few of us can do, at least at first. Also, problem solving is an aesthetic discipline—in other words, an art—where we create and contemplate objects of elegance and beauty. I hope that you enjoy learning about this wonderful subject as much as I have!

Lecture Thirteen

Recasting Integers with Counting and Series

Scope:

This lecture employs the powerful strategies of recasting and rule breaking to 2 classical theorems in number theory: Euler's proof of the infinitude of primes and Fermat's little theorem. We use the knowledge of modular arithmetic and infinite series that we developed earlier. We begin by using simple counting ideas to explore number theory, using the basic principle that if something can be counted, it is an integer.

Outline

I. Let's go back, for a moment, to the earlier lecture about chicken nuggets. If you were really, really observant, you may have noticed a gap in our argument. We associated each lattice point with either a feasible or nonfeasible number. But we blithely assumed for each number n that the equation $ax + by = n$ must pass through lattice points.

- **A.** How do we know if it does?
- **B.** In other words, given relatively prime numbers a and b and an integer n, can we be guaranteed that there are integers x and y such that $ax + by = n$?
- **C.** We only need to consider $n = 1$. If we can find a solution for that, we can easily get one for $n = 2$ and so on.
- **D.** A concrete example: Show that $5x + 7y = 1$ has integer solutions. We will find a result that can easily be generalized to any values of a and b.
- **E.** The idea: Count lattice points.
  1. For each integer n, the equation $5x + 7y = n$ determines a line with slope $-5/7$.
  2. As n ranges from 0 to 35, we get a family of 36 parallel lines.
  3. Notice that if a line intersects one of the lattice points, it will not intersect another. So each lattice point corresponds to a different line.

4. Thus, we just need to count the lattice points.
5. They are arranged in a parallelogram, but we can move the bottom points up. By symmetry, we get a rectangle, and clearly there are 7×5 lattice points.

F. Each of the 35 lines for $n = 0$ to $n = 34$ hits a lattice point. So the $n = 1$ line hits a lattice point, and we are done.

II. We can exploit the idea of counting in many other ways.

III. Combinatorics has its own logic and rules, which are pretty simple, at the start.

IV. Fermat's little theorem.

A. We begin with an example: Find the remainder when 2^{1000} is divided by 13. We can use modular arithmetic to solve this.

1. Start with $2^1 = 2$ (mod 13), and successively multiply by 2.
2. Thus, $2^2 = 4$, $2^3 = 8$, and $2^4 = 16$, which equals 3 (mod 13).
3. Then $2^5 = 6$ and $2^6 = 12$ (mod 13). Now notice that $12 = -1$ (mod 13). So instead of multiplying by 2, we square both sides.
4. We get $(2^6)^2 = (-1)^2 = 1$ (mod 13). In other words, $2^{12} = 1$ (mod 13).
5. This is the crux move, since now we can raise this to any power we want with ease. Since $1000 = 12(83) + 4$, we have $(2^{12})^{83} = 1^{83} = 1$ (mod 13).
6. Hence $2^{996} = 1$ (mod 13), and finally, $2^{1000} =$ 3 (mod 13).

B. The crux: finding the exponent of 2 that equals 1 (mod 13).

C. On the other hand, if we try a nonprime mod, like (mod 10), we discover that the powers of some numbers are never equal to 1 (mod 10), and for others we get 1, but never when we raise to the ninth power.

D. The sensible conjecture to try is that if p is prime, then $a^{p-1} = 1 \pmod{p}$ for any number that is not a multiple of p itself. This is Fermat's little theorem. We will prove this, but first we modify the statement by multiplying both sides by a: $a^p = a \pmod{p}$.

V. Let's replace number theory with combinatorics. We will prove Fermat's little theorem for the concrete values $p = 7$ and $a = 4$ and demonstrate that $4^7 - 4$ is a multiple of 7 using simple counting principles.

A. Imagine a necklace with 7 identical beads. We wish to color them using any of 4 colors. How many different necklaces are possible?

B. Let's make it easier. If it were not a necklace but just a line of beads, then the number of different necklaces would be $4 \times 4 \times 4 \times 4 \times 4 \times 4 \times 4 = 4^7$. That is promising, since it is a number we are interested in.

C. But since we have a necklace, we can slide beads around.

1. In fact, there will be 7 different linear sequences of colors that are all really the same necklace.
2. Indeed, almost any linear color sequence is 1 of 7 "sisters" that form the same necklace. In other words, each linear sequence is in a 7-member sorority.
3. The only exceptions are the 4 monochromatic sequences. These 4 sequences belong to exclusive sororities: Each sorority has just 1 member.
4. Notice that we are using the fact that 7 is prime. If we had a 6-bead necklace, then the pattern black-red-black-red-black-red would only give rise to 2 sisters.

D. We started with 4^7 linear color sequences.

1. Of these, just 4 were monochromatic.
2. The remaining $4^7 - 4$ sequences can be grouped into 7-member sororities where each sorority member is actually the same necklace.
3. So the total number of different necklaces is $(4^7 - 4)/7 + 4$.

E. So $4^7 - 4$ had to be a multiple of 7, which was what we wanted to prove!

VI. For our final example, we will give a second proof of the infinitude of primes, due to Euler, that is notable for its surprise use of infinite series.

A. Start with the harmonic series, which is infinite.

1. Define $S_k = 1 + \frac{1}{k} + \frac{1}{k^2} + \cdots$, and consider the infinite product $S_2 S_3 S_5 S_7 \cdots$, where the subscripts run through the prime numbers.

2. The infinite product begins with $(1 + \frac{1}{2} + \frac{1}{2^2} + \cdots)$ $(1 + \frac{1}{3} + \frac{1}{3^2} + \cdots)$ $(1 + \frac{1}{5} + \frac{1}{5^2} + \cdots) \cdots$.

B. When we expand this infinite product, it will give us each term of the harmonic series; hence it is infinite.

C. Each S_k is just an infinite (but convergent) geometric series.

D. Now, assume to the contrary that there are only finitely many primes. Then there are only finitely many S_k terms in our product. That would make the harmonic series finite. But it is infinite! So there must be infinitely many S_k terms and hence infinitely many primes!

Suggested Reading:

Vanden Eynden, *Elementary Number Theory.*

Zeitz, *The Art and Craft of Problem Solving*, chap. 7.

Problems:

1. Use Fermat's little theorem to find the remainder when 3^{2009} is divided by 19.
2. How many different necklaces can be made using 6 beads, if each bead is a different color?

Lecture Thirteen—Transcript

Recasting Integers with Counting and Series

In the last lecture, we recast the integers geometrically, primarily by looking at lattice points. In this lecture, we will continue to make discoveries about integers, but instead we will use simple counting ideas at first—combinatorics, which is what mathematicians like to call it. Then later we'll use infinite series. Like the last lecture, what we're presenting here is pretty difficult mathematics. We'll prove Fermat's little theorem, which is one of the most important theorems of elementary number theory, but we're going to prove it in a way that was not found in most number theory textbooks. The method that I'm going to present to you is something I discovered when I was training the International Math Olympiad team in their summer training program at the Naval Academy in Annapolis. I was studying the old Hungarian problems that started some of the first math circles back in the early 1900s. In those books, luckily translated into English, I found a wonderful proof of this well-known number theory fact using methods that I had never dreamed were possible. Our methods, therefore, are problem-solver's methods. They're non-standard ideas, and they don't just shed light on the subject of number theory itself, but they're ideas that should inspire you to apply them in many other contexts.

First let's go back, for a moment, to the earlier lecture about chicken nuggets. If you were really, really observant, you may have noticed a gap in our argument. We associated each lattice point with either a feasible or a non-feasible number, but we blithely assumed, for each number n, that the equation $ax + by = n$ must actually pass through lattice points. How do we know if it does? We know that if one of those lines hits a single lattice point, it will hit infinitely many equally spaced along those vectors. That much we know, but how do we know that it even hits a single lattice point? In other words, given relatively prime numbers a and b and an integer n, can we be guaranteed that there are integers x and y, possibly negative, such that $ax + by = n$? Let's reduce the question a bit. We only need to worry about $n = 1$ because if we could find a solution for $n = 1$, we could easily get one for $n = 2$ just by doubling our answer. Let's do a concrete example. Let's show that $5x + 7y = 1$ indeed has an integer solutions. Of course, it's easy to come up with one just on your own: $x = 3,\ y = -2$ fits the bill just perfectly, but what we'll do is something that can be generalized to any values of a and b, not just $a = 5$ and $b = 7$, as long as a and b are relatively prime. We'll solve the problem by counting. What will we count? Lattice points, of course. For each integer n, the equation $5x + 7y = n$, as you know, it determines a line with a slope of $-5/7$. As n ranges from 0 to 35, we will get a family of 36 parallel lines. Here's a picture of just a few of them. The top one, which goes through lattice points at the right and on the left at (7, 0) and (0, 5), that corresponds to $n = 35$, and the bottom one that goes through the origin will correspond to $n = 0$ since, after all, (0, 0) is plugged in. Then there will be lots of other lines. The dotted line near the top is corresponding to $n = 26$.

Notice that if a line intersects one of the lattice points, it will not intersect another because in our picture, the lattice points are close enough spaced that the vector length can't reach any other lattice points in our zone, just in the parallelogram we've drawn. Each lattice point will correspond to a different line. Therefore, all we need to do is count the lattice points. They're arranged in a parallelogram at present, but we can move those blue points up without any difficulty. By symmetry, we'll get a rectangle, and it's easy to count the number of lattice points there. There are exactly 35, 7×5 lattice points. Thus, each of those 35 lines for $n = 0$ to $n = 34$ hits a lattice point. The one for $n = 35$ hits 2 of them. That one is at the boundary; we're not using that one, but among others, the $n = 1$ line has to hit a lattice point because there are 35 values, 35 lattice points, a one-to-one correspondence. We're done. What we did was fairly subtle, and it required our hard-earned understanding of lattice points from earlier lectures, but the key thing to realize is that we were able to show that an equation had a solution merely by counting something. The key idea is counting.

What we'll do now is apply more elaborate combinatorial ideas than just plain old counting. Here's a simple number theory phenomenon that's easy to verify. If you take any integer n and multiply it by the next integer, $n + 1$, the product is always even. Why is that true? That's not a hard thing to figure out. There are lots of ways of thinking about it. One way is to realize, well, you have 2 consecutive numbers, so one of them is going to have to be even. One of the numbers in the product is going to be even, so the product will be an even number. The natural thing to do now is to try to generalize that phenomenon. What about $n(n + 1)(n + 2)$? When I was in high school, when I was just beginning my carrier as a mathlete, I would go, as I told you earlier, to school at zero period for practice. One of the investigations that a group of us did was with this question of consecutive numbers. We wondered: What can you say about a product of consecutive numbers? We know they would always be even, but if you multiplied 3 of them, would they always be a multiple of 3? Could you do better? We found lots of interesting things, and we had lots of conjectures. We tried to prove them, and the problem got more and more complicated. It wasn't until I was quite a bit older that I realized that we were doing number theory when we should have been doing combinatorics.

Here's the main reason why combinatorics, the science of counting, and number theory, the study of integers, are related but not the same. You might think, well, integers are just numbers. You count them. Combinatorics is just counting. Aren't they the same? Not quite. The basic idea is this: If you can count it, it's an integer. That's a penultimate step. Combinatorics has its own logic and its own rules, and they're pretty simple. They're algorithms for counting more and more complicated things. If we can take a combinatorial entity, something that we can count, we know it's going to be an integer, and conversely, if we take some algebraic entity that we're not sure what it is but show that it's a combinatorial entity, then we've shown it's an integer. In combinatorics, the simplest operation is multiplication, and you know this from experience when you order things off a menu. For example, if you go to a restaurant and there are, say, 4 kinds of soups and 3 kinds of entrée, then there will be 4×3, or 12, kinds of dinners you can order.

You've also probably learned about factorials, which is just an extension of this multiplication process. For example, suppose you have 3 different books along a row, and you want to figure out how many different ways you can arrange the books on the row. You could start with the third book, and you have 3 choices for which book it would be; and then the second book, there are only 2 choices; and the first book, there's only one choice. There are $3 \times 2 \times 1$ ways of arranging those books. We write that number as $3!$, and it's equal to 6. With 4 books, it would be $4 \times 3 \times 2 \times 1$, which is $4!$, or 24. Multiplication is very simple in a combinatorial sense. You just reduce your problem to a menu, and you're set.

There's another operation that's the second easiest. After multiplication comes division. For example, suppose there are 10 people in a room, and they all shake hands, sort of like my old handshake problem but more politely. We don't have anyone holding out. Everybody shakes hands with everybody else among the 10 people. How many handshakes occurred? Let's look at it from my point of view. How many people did I shake hands with? I shook hands with 9 people, and there are 10 people like me with that point of view. That should give us an answer of 10×9, which is 90, but we've overcounted in a uniform way. For example, if Barack Obama was at my party and I shook his hand, I would count that as one handshake, but of course, Barack Obama would count shaking hands with me as one handshake, but it would be the same handshake. What's happening is a uniform overcounting by a factor of 2, so we have to take that 9×10 and divide it by 2 and get the correct answer of 45.

Now when I teach combinatorics to my college students, I like to play a game with them called Combinatorial Jeopardy, where I write down a mathematical expression, some computation, and I ask them to give me a question for which my expression is the proper answer. For example, if we start with the answer of $10 \times 9 \times 8$, what's the question? One question would be: How many ways could you pick a 3-person committee with a president, a vice president, and a treasurer out of a 10-person pool? You would say to yourself, "Okay, who's going to be president?" And there'll be 10 choices. It's a menu. Then, who's going to be vice president? That would be your soup maybe. There are 9 choices now because you chose someone already, and likewise, there'll be 8 choices for treasurer. The answer would be $10 \times 9 \times 8$, simple multiplication.

Now, what if we didn't care about the offices in this committee? We didn't care who was president. In fact, there wouldn't be any. It's just a committee of 3 people. Then if we picked Andy, Betty, and Carlos, it's the same as if we picked Betty, Andy, and Carlos. The order that we picked them doesn't matter. There's going to be uniform overcounting, but by what factor? Well, there's Andy, Betty, Carlos; Betty, Andy, Carlos; Carlos, Betty, Andy; etcetera. How many ways can we arrange those 3 people? That's the same as how many ways can we arrange 3 books in a row, and the answer would be 3!, which is 6. We have uniform overcounting by a factor of 6. In other words, what we've determined is that $10 \times 9 \times 8$ divided by 6 counted a combinatorial entity. So $10 \times 9 \times 8/6$ is an integer. In other words, putting this into the number theory context, the product of

$10 \times 9 \times 8$ is divisible by 6. It would work in general. The product of any 3 consecutive numbers is always a multiple of 6. By the same logic, the product of 4 consecutive numbers could be divided not just by 4 but by 4!. Therefore, the product of 4 consecutive numbers is automatically a multiple of 24, and in general, the product of k consecutive powers is a multiple of $k!$.

What we've managed to do is prove an interesting number theory fact using nothing but counting, just using the idea that if you can count it, it's an integer. We're going to extend this method and use it to prove something more important, namely, Fermat's little theorem. What Fermat's little theorem is about is a way of predicting regularity in the patterns with modular arithmetic of exponents. For example, if you look at the powers of 2, you get 2, 4, 8, 16, 32, and so on, and if you just look at the rightmost digit, you get 2, 4, 8, 6, 2, 4, 8, 6. We have a pattern that keeps repeating. When you look at the rightmost digit of a number, what you're really doing is just saying: What is that number modulo 10? Modulo 10 the only exponents are 2, 4, 6, 8. The powers of 5, the only things they are equal to is 5—5, 25, 125; they're all congruent to 5 modulo 10. The same thing with powers of 6 modulo 10. Modulo 10 is kind of boring. The reason modulo 10 is kind of boring is 10, unfortunately, is not a prime, and remember, the primes are the fundamental objects in number theory. They're sort of the pure things; you can't break them down. Fermat's little theorem, it's proper venue if investigation is exponents modulo a prime.

Let's look at an example. This was a problem that I remember first learning about in high school, and it really hooked me into wanting to learn more about number theory. It was from one of the very first of these math contests, one of the feeder exams that would lead towards the USA Olympiad and the International Math Olympiad. The question was: Find the remainder when 2^{1000} is divided by 13. Let's do it with modular arithmetic, and we'll just move ourselves into the mod 13 universe, where there are only 13 different numbers. We'll start out with 2^1. That's equal to 2, and $2 \equiv 2$ modulo 13. Everything is congruent to itself modulo 13. Now let's keep multiplying by 2. So $2^2 \equiv 4$, $2^3 \equiv 8$, and $2^4 \equiv 16$, which in turn is congruent to 3, reducing it. So $2^4 \equiv 3$. This is the power of mods. We don't have to do a hard calculation. Instead of dealing with the number 16, we just have 3 to kick around. So now 2^5, we're doubling once more, but we're not going to double 16. We just double 3;

$3 \times 2 = 6$, so $2^5 \equiv 6 \pmod{13}$. Likewise, $2^6 \equiv 12 \pmod{13}$. Now we have a breakthrough; 12 and 13 are so close and we are so lazy about calculation, why multiply 12 by 2? That's a lot of work; 12 and 13 are so close, we will replace 12 with $-1 \pmod{13}$. After all, $12 \equiv -1 \pmod{13}$. It's another way of saying that 12 is 1 less than a multiple of 13, or $12 - -1$ is a multiple of 13.

Now we have $2^6 \equiv -1$. What should we do with that? Should we keep doubling it? No. Let's square both sides. The reason is if you square -1, you get 1, and if you square 2^6, what do we get? We get 2^{12}. We've suddenly jumped all the way up and found that $2^{12} \equiv 1 \pmod{13}$. Why is that useful? What's important about that? It's our crux move, in a way, because once we know something is congruent to $1 \pmod{13}$, the pattern starts anew: $2^{13} \equiv 2$, so it will be just like 2^1, and then 2^{14} will be just like 2^2, and 2^{15} will be just like 2^3, and so on. What we know is that the exponents modulo 13, the exponents of 2, have a 12-step repeat pattern that ends with 1 at the 12th power; it ends with 1 at the 24th power, and so on. Every 12th power will be congruent to 1. Let's keep our eyes on the prize, which was to figure out 2^{1000}. Can we get to 2^{1000} quickly now? Yes, we just raise 2^{12} to a high enough power to get right near there. It's sort of like we're going into hyperspace; we just jump all the way up to 2^{996} by raising 2^{12} to the 83rd power, and we can do it instantly because all we're really doing is raising 1 to the 83rd power. We get $2^{996} \equiv 1 \pmod{13}$, and finally, we can get to 2^{1000}. It's just 4 more powers of 2. It's $(2^{996})(2^4)$, which is the same as (1)(3) because remember $2^4 \equiv 3$, so our answer is 1×3, which is 3. In other words, if you took 2^{1000}, which is a 302-digit number, I believe, and divided it by 13, you would get some horrible quotient and a remainder of 3.

You might not think that's an important problem, and, in fact, it's not an important problem. There's no useful reason to know that the remainder of 2^{1000} when you divide it by 13 is 3. This is not something you should memorize, but the idea that there are these patterns with exponents, that's interesting because it can be generalized. If you notice it was 2^{12} was congruent to $1 \pmod{13}$. You might want to conjecture something. Maybe 3^{12} would also be congruent to $1 \pmod{13}$, and indeed, it is. In fact, if we try just one other example and try a different mod. Instead of mod 13, we would discover that if we tried to look at 4^{10} modulo 11, we get $4^1 = 4$, $4^2 = 16$, which is the same as 5, and if we keep on going, indeed, we'll discover that $4^{10} \equiv 1 \pmod{11}$. If we did this with a non-prime like 10, as we know, we're not going to get any interesting patterns, so we'll stick to primes.

Let's make a bold leap and make this conjecture, which is Fermat's little theorem, which says that a, in general, if a is any number that is not a multiple of p, $a^{(p-1)}$ would be congruent to 1 (mod p). That's called Fermat's little theorem to contrast it from Fermat's last theorem, which was Fermat's famous conjecture proven only at the end of the 20th century. Fermat's little theorem, again, says that if a is not a multiple of p, then $a^{(p-1)}$ is congruent to 1 (mod p). We'll prove it, but first, let's modify it a tiny bit by multiplying both sides by a to make it just a little prettier. So we get $a^p \equiv a \pmod{p}$. Note that this will be true even if a is a multiple of p. We have a nice, sort of simple, generic equation: $a^p \equiv a \pmod{p}$. That's what we'd like to prove. Again, if we go back to mod 13, 2^{13} now would be congruent to 2 (mod 13). Can we prove something like this in general?

Fermat's little theorem is a very, very standard important theorem of basic number theory, and there are many standard proofs for it. We're going to do it in a completely different way. We're going to take the assertion that $a^p \equiv a \pmod{p}$and replace that with the phrase $a^p - a$ is multiple of p. Now we do some recasting using our combinatorial ideas. Let's leave the number theoretic world behind and go back to counting. Let's look at a concrete case where $a = 4$ and p is the prime 7, and we'll prove without calculation that $4^7 - 4$ must be a multiple of 7. What we're going to do is count necklaces. Imagine a necklace with 7 identical beads. We wish to color them using any of 4 colors. How many different necklaces are possible? Let's make it easier first. If it were not a necklace but just a line of beads, then the number of different necklaces would be $4 \times 4 \times 4 \times 4 \times 4 \times 4 \times 4 = 4^7$. This is promising because it's a number we're interested in, but since we have a necklace, we can slide beads around. For example, here are 2 different color sequences that actually form the same necklace. We just slid the right one over to the left and pushed everything over. In fact, there will be 7 different linear sequences of colors that are all really the same necklace. We're overcounting in some sense. Indeed, any linear color sequence, with a few exceptions, is one of 7 "sisters" that form the same necklace. In other words, each linear sequence, you can think that it belongs to a 7-member sorority. The only exceptions are the 4 monochromatic sequences, such as black-black-black-black-

black-black-black. These 4 sequences with the 4 different colors belong to exclusive sororities; each sorority has just one member. We're using the fact that 7 is prime here. If we had a 6-bead necklace, then the pattern black-red-black-red-black-red would give rise to 2 different "sisters," not 6 different "sisters." Likewise, the pattern black-red-green-black-red-green would give rise to 3 different "sisters."

Getting back to our problem, what we started with were 4^7 linear color sequences. Of these, just 4 of them were monochrome. The remaining $4^7 - 4$ sequences—we're getting rid of the monochrome, so what's left is $4^7 - 4$, which is the number we wanted to think about—of that number of sequences, they can all be grouped into 7-member "sororities," where each sorority member is actually the same necklace because we overcounted uniformly by 7. The total number of different necklaces would be $(4^7 - 4)$, divided by 7, the size of the sororities, plus 4, which is really counting 4 exclusive sororities. That's the number of necklaces. Yeah, we counted necklaces, but we didn't really care about necklaces. If you look at what we did here, what we did was we counted something, something combinatorial. We counted it, so it's an integer, and therefore, $4^7 - 4$ when we divided it by 7, that's an integer. That tells you that $4^7 - 4$ is a multiple of 7. We sneakily were able to get a congruence equation just by counting something. You can see this method would be completely general. It would work for any prime p and any integer a; just consider a p-bead necklace that uses a different colors. We were able to prove a beautiful number theory result by recasting it as a combinatorial problem. It's a little known but very powerful technique.

We're going to conclude now by looking back at the infinitude of primes. We already proved that using that using the ancient Greek method that was popularized by Euclid, but I want to show you yet another way of doing it that was discovered by the great mathematician Euler in the 18th century because the method uses a beautiful example of chainsawing the giraffe. It's not a combinatorial method now, but we use infinite series. What Euler did was to start with the harmonic series $1 + 1/2 + 1/3 + 1/4 + 1/5$, which you know diverges; remember that blows up to infinity. We know that that's not a sum that converges to a finite value. And his bold idea was to factor it as if it were a polynomial. That's not really legal. You can't play with infinity as if it's a number, but Euler didn't care. Euler was able to chainsaw giraffes before the chainsaw was invented. Euler took the harmonic series—think of it as being equal to infinity—and he tried to factor it. The way he factored it was he looked at infinite sums, not the harmonic series, but infinite sums of the reciprocals of primes. For example, if we would define S_2 to be $1 + 1/2 + 1/2^2 + 1/3^2$ and so on at that infinite sum, and S_3 to be the same thing but for the reciprocals of the powers of 3, and S_5 to be the sum $1 + 1/5 + 1/5^2$ and so on, if you look at the product $S_2S_3S_5S_7S_{11}S_{13}$, going through all the primes, what you'll get is a product of infinite series.

The first few terms look like this: $(1 + 1/2 + 1/4 \ldots)$ $(1 + 1/3 + 1/9 + \ldots)(1 + 1/5 + 1/25 + \ldots)$. If you expand it—just pretend you're in algebra 1 and you're expanding polynomials, only its really scary and infinite, but just get started, [and] you'll notice the first few terms are $1 + 1/2 + 1/3 + 1/4 + 1/5 + 1/6$. In other words, it looks like we're getting the harmonic series. Why? If you think about it, for each number n, the term $1/n$ will appear in our factorization. For example, suppose we have a giant number like $n = 360$, then $1/360$ will appear when we multiply the term $(1/2^3)(1/3^2)(1/5)$ because 360 factors into $(2^3)(3^2)(5)$. Every number factors into primes, and so the reciprocals of those appropriate powers of primes will be in those infinite sums, and there will be only one case of doing that. It's unique because factorization into primes is unique, so for each number, there's a unique factorization into primes, and we'll find them lurking about as reciprocals in those infinite sums. Every $1/n$ will appear when we multiply those infinite sums, which is pretty wild.

What does this have to do with the infinitude of primes? If you think about it, each of those infinite sums, like, say, S_7 is $1 + 1/7 + 1/7^2 + 1/7^3$ and so on, that's an infinite geometric series, and we learned earlier how to sum them up. There's a formula for it, which you've probably forgotten. That doesn't matter. The important thing is the value will be finite as long as the common ratio is less than 1, and here it certainly is. The sum will be finite. So each of those S's, S_2 $S_3S_5S_7$ and so on, each of them is just a number, a finite number. What Euler showed is that the harmonic series is equal to a product of these $S_2S_3S_5S_7 \ldots$ Now, assume to the contrary, that there are only finitely many primes, and there are only finitely many S_k's in our product. Each of them is finite, so we'll have a finite product of finite numbers and we'll get a finite result, and then the harmonic series would be finite. But it's not finite; there's our contradiction. The only way out of it is that there have to be infinitely many primes—absolutely wild, unexpected. Who would have thought that infinite series could teach you about primes? It turns out that Euler's little trick here led to an entire branch of mathematics called analytic number theory, which is the application of the science of infinite sums, which is calculus essentially, to number theory.

The power of simple counting, plus these more elaborate arguments, based on the idea that if you can count something it's an integer, are incredibly productive. Then this Eulerian chainsawing that we've seen, it's also an incredibly impressive, productive idea. Let's step back from it for a second because you might be saying, "Wow, that's impressive, but how can I emulate it? I'm not going to invent analytic number theory." There are 2 things to take away from looking at what Euler did. First of all, there's the mathematics itself, which is very interesting and something that leads in lots of directions. Just remember that you have my permission and, certainly, Euler's permission to practice creative theft here. The other thing is the importance of boldly breaking a rule. In other words, chainsawing, not in a stupid, random way, but more in a sort of a boundary-stretching way. We're not talking about a paradigm shift. What we're talking about is something more modest. Remember, in things like art, people will complain and say, "Well, a person can throw junk against a wall and call it art." You can't quite do that in mathematics. You can't randomly just do weird things because in mathematics something is either right or something is wrong. But I don't want you to be afraid of doing weird things because something to remember also about mathematics is the importance of failure. If most of your attempts are successful, then you're not attempting anything really interesting. On the other hand, if 90% of your attempts are failing, then you're probably doing some interesting mathematics. Please remember that. Experiment, have fun, and don't worry about failing. Failing will only teach you more math.

Lecture Fourteen

Things in Categories—The Pigeonhole Tactic

Scope:

Like the extreme principle, this tactic seems almost vacuous: If you try to put $n + 1$ pigeons into n pigeonholes, at least 1 hole will contain at least 2 pigeons. Yet the pigeonhole principle allows us to solve an amazing variety of problems. Among the applications we will explore is a graph theory subject known as Ramsey theory, which provides a systematic way of finding patterns in seemingly random structures.

Outline

I. The simplest version of the pigeonhole principle: If you have more things (pigeons) than categories (pigeonholes), at least 2 of the things belong to the same category.

A. Example: Suppose you color the infinite plane in 2 colors, red and blue, in any arbitrary way. Prove that there are 2 points exactly 1 meter apart that are the same color.

1. Pigeonhole solution: Consider an equilateral triangle with side length 1 meter.
2. Each of the 3 vertices is colored; there are only 2 possible colors.
3. Let the vertices be pigeons, and the colors be holes. Since $3 > 2$, there is a hole with at least 2 pigeons.

B. Example: People are seated around a circular table at a restaurant. The food is placed on a lazy Susan in the center of the table. Each person ordered a different dish, and it turns out that no one has the correct dish in front of him or her! Show that it is possible to rotate the platform so that at least 2 people will have the correct dish.

1. People are the pigeons, but what are the holes?
2. This is the challenging part of using the pigeonhole principle: carefully formulating a penultimate step that will solve the problem.

3. Suppose there are n people. We can measure distance around the table where n units is a full circle. Everyone starts out a certain nonzero clockwise distance from their correct dish. The possible distances are thus 1, 2, 3, … , $n - 1$.
4. So at least 2 of them are the same distance from their correct dish.
5. Move that distance, and we are done!

C. Example: Prove that among any group of people, 2 of them have the same number of friends in the group.

1. Crux recasting: graph theory! Prove that in any graph, 2 of the vertices must have the same degree.
2. This problem seems perfect for the pigeonhole principle, with vertices as pigeons (things) and degree values as holes (categories). For example, in a 6-vertex graph, there are 6 pigeons. So our penultimate step would be 5 possible degree values.
3. But there are 6 possible values! In a 6-vertex graph, a vertex can have 0, 1, 2, 3, 4, or 5 neighbors.
4. Wishful thinking and the extreme principle tell us that there must be a way to get rid of at least 1 degree value.
5. Consider degree 0. If a graph has such a vertex, it is isolated from the others, in which case no vertex can have degree 5.
6. Conversely, if a vertex has degree 5, it is connected to all other vertices, so no vertex can have degree 0.
7. Thus the possible degree values are between 0 and 5, inclusive, but cannot include both 0 and 5. So there are only 5 possible values, and with 6 vertices, we can conclude that 2 vertices must have the same degree!

II. The intermediate version of the pigeonhole principle: Suppose we put p pigeons into h holes.

A. Then at least 1 hole contains at least $\lceil p/h \rceil$ pigeons, where the brackets mean ceiling (the ceiling of x is the least integer that is greater than or equal to x).

B. Example: Suppose you have a drawer with 23 socks, and the socks come in 4 colors. Then you must have $\lceil 23/4 \rceil = \lceil 5.75 \rceil = 6$ socks that are the same color.

III. Ramsey theory: Named after Frank Ramsey, this is a branch of discrete math that concerns itself with what kind of order is guaranteed, even in a random structure.

A. A typical Ramsey theorem says something like, "Given a large enough structure, we are guaranteed to see a smaller substructure."

B. For example, no matter how we 2-color the plane, we are guaranteed 2 points a meter apart that are the same color.

C. Here is a classic example: Show that among any 6 people, either 3 of them are mutual friends, or 3 are mutual strangers.

1. Graph theory recasting: If you 2-color the edges of a complete graph with 6 vertices, then there must be a monochromatic triangle!
2. Use green and red. We want to see lots of one color, since that would make it more likely to get a monochrome triangle. The extreme principle suggests that we search for the vertex that has, say, the maximum number of red edges emanating from it.
3. The pigeonhole principle gives us that maximum: Each vertex has 5 edges emanating from it. At least $\lceil 5/2 \rceil = 3$ edges are the same color (say, red).
4. Focus on the vertex we started with plus the 3 others that are joined to it with a red edge.
5. If any of these 3 vertices are joined with a red edge, we are done.
6. But if none of them are red, we have created a green monochromatic triangle.

IV. Let's generalize this problem. We just showed that if the edges of a K_6 are 2-colored, then there must be a monochromatic triangle. The Ramsey number formulation of this is R(3, 3) = 6.

A. The Ramsey number R(*a*, *b*) is defined to be the smallest number *N* such that if the edges of a K_N are colored blue and red, then there must be a red K_a or a blue K_b.

B. Ramsey numbers can use more than 2 colors. For example, R(3, 3, 3) is equal to the smallest *N* such that if you 3-color the edges of a K_N, you must have a monochromatic triangle.

C. Ramsey's theorem states that the numbers R(a, b, c, …) exist and are finite.

D. Example: R(3, 3, 3), the only nontrivial Ramsey number known involving more than 2 colors, is equal to 17.

Suggested Reading:

Soifer, *Mathematics as Problem Solving.*

Zeitz, *The Art and Craft of Problem Solving*, sec. 3.3.

Problems:

1. Given a unit square, show that if 5 points are placed anywhere inside or on this square, then 2 of them must be at most $\frac{\sqrt{2}}{2}$ units apart.
2. People have at most 150,000 hairs on their head. How many people must live in a city in order to guarantee that at least 10 people have exactly the same number of hairs on their head?

Lecture Fourteen—Transcript

Things in Categories—The Pigeonhole Tactic

In this lecture, we're going to look at the famous pigeonhole principle, which is another one of the sort of magical tactics, like the extreme principle, that can solve hard problems almost instantly. The pigeonhole principle has a narrower field of applicability compared to the extreme principle, but it's still a very, very powerful idea. The pigeonhole principle is very, very common knowledge among young mathletes. It's often, however, their first and only higher-level problem-solving idea. The pigeonhole principle has been glorified in American math circle/math team culture, but it's really just one of many important tactics. In Europe, it's not called the pigeonhole principle; it's usually called the Dirichlet principle, named after Dirichlet, who was a German mathematician with a French name, who used it to do very important discoveries in analytic number theory, applying calculus to the theory of numbers. The pigeonhole principle stems from a logical penultimate step that's often used to get nearly to the end of many problems. The reason it's useful is it's secretly a way of getting a penultimate step to work. You'll see what I mean in just a moment. What we'll do in this lecture is use the pigeonhole principle in a few different mathematical fields, in geometry and number theory, and then we'll extend it to move forward into an area of graph theory known as Ramsey theory. That's a branch of math that allows us to predict the existence of orderly patterns in seemingly random systems. We'll apply our problem-solving ideas to some very important, practically philosophical problems in mathematics.

What is the pigeonhole principle? It's very simple. It says if you put p pigeons into h pigeonholes and $p > h$, then at least one of the holes will contain at least 2 pigeons. In other words, if you have more people than seats and you're playing musical chairs, instead of somebody getting kicked off, somebody is going to sit on someone's lap, a completely obvious fact. The way to think about it is: If you have more things than you have categories and you have to put the things into categories, you're going to have to have at least 2 things with the same category. The pigeonhole principle is a logical idea relating things to categories. I'll often call the things pigeons and the categories holes, but you can use any analogy you want. I like pigeons and holes because it's just kind of fun thinking about animals, but that's up to you.

Let's take a look at it and see why it's useful. Here's a problem in a branch of math called combinatorial geometry. It's the kinds of counting questions you get when you look at things like the plane. We've already done little bits of combinatorial geometry with our lattice point counts. This is a much simpler question. Suppose you color the infinite plane just using 2 colors, red and blue, and you do it in any old way, possibly an amazingly complicated way, like a fractal painting with reds and blues. Perhaps every point where the coordinates are rational, you paint blue, and if the coordinates are irrational, you paint them red. Something like that, but you just do a random painting. You're handed this random painting of the entire plane. What we will prove is that no matter how it was painted, there are 2 points exactly 1 meter apart that are the same color. Maybe I shouldn't use the word "paint" because when you think of painting, you think of a brush. Think of this as mathematical painting, where each point on the plane is assigned a color. Don't think of it as a physical thing where there are blobs of paint. You can't use blobs of paint. You have no control over what color any point is; it's just red or blue, but you do have the pigeonhole principle. Let's think of something involving categories and things. We have 2 colors, and the pigeonhole principle requires something to be bigger than something else. Let's think of 3 things. How about 3 points? Let's put them on an equilateral triangle and give them a side length of 1 meter. Consider an equilateral triangle in the plane where the vertices are 1 meter apart. We have 3 points. Those are our things; those are our pigeons. We have only 2 colors, 2 categories, in which to place them. You've seen this argument before, without loss of generality, 2 of them must be red, and we've solved our problem. We have forced the existence of 2 points exactly 1 meter apart that are the same color. That was a pretty easy problem, and there are other ways of solving it, but the pigeonhole method is nice and fast. The pigeonhole method is fun, too. That's another reason why the pigeonhole principle has been glorified and definitely overemphasized in the mathlete set. But it is a lot of fun to play with the pigeonhole principle; it's even fun to say pigeonhole. I don't know about Dirichlet, but saying pigeonhole is fun.

What we did is a very simple problem about coloring, but it actually gets us close to the research frontier. If we replace 2 colors with 3, or 4, or 5, the question becomes much harder, and there have been many papers and books written about it. One of the most recent ones that came out is entitled *A Mathematical Coloring Book*, which was written by a friend of mine, one of my mentors, who was a fairly early immigrant from the Soviet Union. He started, I think, the first regional Olympiad in this country, in Colorado Springs, back in the 1980s. When we're using the pigeonhole principle, the difficulty is picking what are your things and what are your categories. You need to be quite flexible about that. They could end up being the opposite of what you think.

Let's do another example to give you a feel for it. This is a story that I actually heard at a Chinese banquet at a math conference, and it was inspired by the fact that we were sitting at a banquet table where the center of the table can rotate. Here's the problem: Imagine people are seated around a circular table at a restaurant, and the food is placed on a circular platform in the center of the table, and the circular platform can rotate. Each person ordered a different dish, and it turns out that no one has the correct dish in front of them. Show that it's possible to rotate the platform so that, in fact, at least 2 people will have the correct dish. We start where no one has their correct dish. The question is: Can we prove that it's possible to rotate the circular platform in the center of the table so that at least 2 people will have their correct dish? How do we get that to happen? What are the pigeons? What are the holes? What are the things? What are the categories? The challenging part is to come up with the penultimate step that will get us to the solution. In this case, we need categories defined so that if 2 people are in the same category, then they both have the correct dish. Notice that the dishes themselves won't work as categories. First of all, we know that there are exactly as many dishes as there are people. Also, it doesn't make sense; each person has a unique dish. We need to use what the problem gives us, and what the problem gives us—visualize it; it gives us motion. The table is spinning.

Now we're on to something; where there's motion, there's time and distance. For this problem, let's look at distance. Suppose there are, say, 10 people. We can measure distance around the table where 10 units is a full circle. Everyone starts out a certain non-zero clockwise distance from their correct dish. The possible distances would be 1, 2, 3, 4, 5, 6, 7, 8, and 9, not 0, because if 0 were one of the distances, then someone would have their dish in front of them. Think of it like this: Each person looks at the food around them and instead of saying, "I don't have my dish," what they say is, "If we only turned the table by 3 units, I'd be happy." If you query the 10 people, at least 2 of them will be saying the same number of units because, after all, there are only 9 different distances people can be away from their food and there are 10 people. The 10 people are the pigeons, and the 9 distances are the holes. We were able to put 2 people into the same category of displacement from food, and we solved the problem.

Let's do another one. Here's an interesting fact: Among any group of people, 2 of them have the same number of friends in the group. That's assuming, of course, that friendship is mutual. If you're friends with me, then I'm friends with you. For example, if we looked at all the people in the world who like spicy food, there are going to be at least 2 of them who have the same number of spicy-food-lovers as friends. How do we look at this problem? How do we investigate it? You know by now any problem involving people and friends, we turn it into graph theory. The question instantly recasts into the useful form: In any graph, 2 of the vertices must have the same degree. Now we can draw pictures. Here's an example of a 6-vertex graph, and you can see several of the vertices have the same degree—a degree of 3; there are degrees of 2, but that, of course, might have been a coincidence. Is it possible to draw a graph where the vertices have different degrees, where all the vertices have different degrees? It seems to be impossible if we tried drawing lots and lots of graphs, but that's not a proof. This problem seems perfect for the pigeonhole principle, with vertices as the pigeons, as the things, in other words, and degree values as the holes, or the categories. For example, in the graph we just examined, there were 6 pigeons—there were 6 things—so our penultimate step would be to find 5 possible degree values, but—whoops!—there are 6 possible degree values. In a 6-vertex graph, a vertex could have 0, 1, 2, 3, 4, or 5 neighbors. What should we do? Maybe we should give up. No, wishful thinking and the extreme principle tell us that there must be a way to get rid of at least one degree value. Consider degree 0. If a graph has such a vertex, it's isolated from the others, in which case, no vertex can have degree 5. Likewise, if a vertex has degree 5, it's connected to all the other vertices, so no vertex can have degree 0. It's a lot like the handshake problem that we did earlier. Thus, the possible degree values are between 0 and 5 inclusive but cannot include both 0 and 5. There are only 5 possible values with 6 vertices, and so we can use the pigeonhole principle to conclude that 2 of those 6 vertices, at least, must have the same degree. Notice that this problem was not that hard. The hardest part was just not giving up when we hit a little bit of a bump on the road.

Let's look at a few number theory problems now. Given 18 integers, can you prove that at least 2 of them have the same remainder upon division by 17? That's easy. That's a simple application of the pigeonhole principle. In fact, pigeonhole principle problems are often written with that sort of language. You see 17 and you see 18 in the same sentence, you instantly know that this sounds like a pigeonhole principle problem. There are only 17 possible remainders (mod 17), namely, 0, 1, 2 through 16. At least 2 of them have the same remainder. In other words, 2 of them are going to be in the same pigeonhole. You should actually try to visualize integers falling into buckets or holes and they're labeled 0, 1, 2, 3 through 16. You have 18 of them falling from the sky. Two of them are going to have to land in the same place. It helps to visualize, remember, and it's fun.

Here's a problem from the Colorado Math Olympiad that the author of the *Mathematical Coloring Book* founded. This was back more than 20 years ago that this problem appeared. Show that if you have 17 positive integers, then either 1 of them is a multiple of 17 or a sum of several of them is a multiple of 17. That's a little bit strange, and 17 is a little too large. Let's replace 17 with 5, and that way we can at least write down numbers. We have the numbers a, b, c, d, and e. What we want to show is that either 1 of them is a multiple of 5 or that some sum of them is going to be a multiple of 5. If you have sums, there are lots of ways of writing them, but let's do it in the most organized, simple way. Let's just consider the following numbers: a, $a+b$, $a+b+c$, $a+b+c+d$, and $a+b+c+d+e$. We have 5 numbers. Now the problem is a modulo 5 problem, but just like the problem with the friends, it's possible that one of these numbers is already congruent to 0. If it's already congruent to 0, then we're done because either the number a or one of the sums is going to be a multiple of 5. On the other hand, if none of them [is] congruent to 0, then we have only the remainders 1 through 4 available, yet we have 5 numbers. Again, by the pigeonhole principle, those 5 numbers are falling into 4 categories, and 2 of them have to be in the same category. For example, $a+b$ and $a+b+c+d$ could have the same remainder. If they have the same remainder, then their difference will have a remainder of 0, and the difference of $a + b$ and $a + b + c + d$ would just be $c + d$. We'd end up—the number $c + d$ would have a remainder of 0. Notice we use the fact that 2 things were in the same category modulo 5 to get a remainder of 0 (mod 5).

Let's move on to an intermediate version of the pigeonhole principle. Suppose we put p pigeons, again, into h holes. Let's think about it a little more. If p is bigger than h, we're guaranteed to have cohabitation of pigeons, but we could find more than that. What we can determine is: If you have p pigeons in h holes, if you divide p by h, you'll get a number, possibly with a remainder. If you take the ceiling of that number, in other words round it up, that will be the number of pigeons that's guaranteed to be in at least one hole. For example, suppose you had a drawer with 23 socks, and the socks come in 4 different colors. If you go 23 divided by 4, you get 5.75, and if you take the ceiling—if you round it up—you get 6. The pigeonhole principle says you're guaranteed to have 6 socks, at least, of the same color—23 socks, 4 colors; it's going to force at least 6 socks to be the same color, and it's easy to see why. It's just proof by contradiction. Suppose, on the contrary, you have just 5 or fewer socks of the same color. Since you have only 4 colors, you'd have, at most, 5×4, which is 20 socks. As soon as you get above 20, you're forced into having 6 of the same color. That's the intermediate pigeonhole principle, so to speak.

This very simple observation allows us to now do some serious mathematics. We're going to start to take a look at the Ramsey theory that I referred to earlier. Ramsey theory is named in honor of Frank Ramsey, who wrote a brief paper about this in the year 1930, which was the same year that he died, at the age of 26. If Ramsey had lived a longer life, he'd probably be one of these mathematicians glorified along with the likes of Gauss or Euler because he truly was a genius, and he was incredible productive, not just in mathematics but in economics and philosophy, too. The Ramsey theory, named after Ramsey, was just, from his point of view, a little throwaway lemma in a paper he was writing about philosophy. Nevertheless, it has given rise to a gigantic branch of mathematics. What a typical Ramsey theorem says is something like this: Given a large enough structure, we're guaranteed to see a smaller substructure. Here's an example: No matter how you color the plane with 2 colors, you're guaranteed to see 2 points a meter apart that are the same color. You have a random coloring of a giant structure. It forces some order into a smaller substructure.

Here is a simple, classic example. Let's show that among any 6 people, either 3 of them are mutual friends or 3 of them are mutual strangers. I want to remind you that this is mutual, so you don't have unrequited friendships. I also want to remind you about the word "or." "Or" does not mean "exclusive or." In the world of mathematics, "or" can always mean both. In the world of ordinary English, "or" often has an exclusivity to it. For example, if you're a mathematician and your significant other says, because you've been working so hard at math, says something like, "You know, it's me or the math," and you just say to them, "Yes." They'll probably hit you, and you won't understand why they hit you because the word "or" could mean both. What we want to show is that among any 6 people, 3 of them might be mutual friends or 3 of them are mutual strangers, and it's possible that you could have both of these occurrences at once. We know to recast this with graph theory. What we want to do is take a 6-vertex graph, and what we want to do then is draw all of the edges of that 6-vertex graph. This is called a complete graph, and it's often abbreviated, if it has 6 vertices, as a K_6, K standing for complete.

Imagine taking a graph with 6 vertices. Draw all of the edges, and then what we want to do is color an edge one color for friendship and another color for strangerness. What we want to prove is that we're going to have a situation of 3 people that are all connected by the same type of relationship. In a visual sense, what we want to prove is that there must be a monochromatic triangle, one with all edges the same color. Let's investigate it. We'll use green and red, and what we want to do is see lots of one color because the more color we see, the more likely it will be that we'll get our monochrome triangle. The extreme principle would dictate that we search for the vertex that had, say, the maximum number of red edges emanating from it. The pigeonhole principle gives us that maximum. Each vertex has 5 edges emanating from it because there are 6 vertices in the graph, and so by the intermediate pigeonhole principle, if we divide 5, the number of things, the number of edges, by 2, the number of colors, we get 2.5. When we take the ceiling of that, we get 3. By the pigeonhole principle, at least 3 edges are the same color, and without loss of generality, let's say that color is red. Three red edges is the best we can do, but it's enough. Now we can focus on the vertex we started with because now that's our natural point of view for our investigation. Then, we'll look at the 3 others that are joined to it with a red edge. If any of those 3 vertices were joined with red edges themselves, then we'd be done. We would have a monochrome red triangle. For example, if edge *bd* were red, then *abd* would be our monochromatic red triangle. What if none of them [is] red? That's possible, but then they're all green, and then we have created, say, the monochromatic triangle *bdc*, so we're done. Either we'll have a red one, or we'll have a green one, possibly, but we're guaranteed to have one or the other. That's the simplest kind of a Ramsey theorem.

Let's generalize the problem. What we've shown is that if you were to color the edges of a K_6, in other words, a graph with 6 vertices where all the edges are joined, and we 2-color it, so to speak—we color it with 2 colors—then there must be a monochromatic triangle. There's a Ramsey formulation of this. We write the expression R(3, 3) = 6; we read it "R of 3, 3 equals 6," or just R(3, 3). It's kind of complicated notation. The way it works is the Ramsey number R(*a*, *b*) is defined to be the smallest size *n*, such that if the edges of a K_n were colored blue and red, then there must be a red K_a—in other words, a K_a that's completely monochrome; all of its edges are red—or a blue K_b. That's a little hard to grasp. Let's try a concrete example, R(5, 7). That would be some number *n*, the smallest possible *n*. It could be really large, say, 1,000,000,000,000, such that if you took a complete graph with 1,000,000,000,000 vertices and joined all the edges, color them blue or red, then it will guarantee the existence of a K_5 somewhere in that graph, where there are 5 vertices all joined and all those vertices are joined by red edges, or a blue K_7, 7 vertices where all the edges joining them are blue. It's a pretty abstract idea. It turns out, though, that the Ramsey numbers can be used. You can take this idea and make it even more complicated. We can use more colors. For example, we can define a number R(3, 3, 3). That's equal to the smallest *n* such that if you 3-color the edges of a K_n, you'll have a monochromatic triangle. The reason it works is its R(3, 3, 3). What the first 3 stands for is a 3-sided figure in red. The next 3 is a 3-sided figure in blue, and the next one is a 3-sided figure in orange, perhaps. So R(3, 3, 3) is the smallest *n* that will guarantee that you'll either have a triangle in one color, or a triangle in another color, or a triangle in a third color.

Ramsey's theorem, which is a grand, grand idea, states that these very abstract, weird entities—the numbers R of *a*, *b*, *c*, any finite number of things—these numbers exist and they're finite. In other words, given any imaginable substructures that you hope to see, you'll see them if you have a large enough n. The reason I'm spending a lot of time on this notation is not to torture you but to tell a story later. Bear with me and wait a few lectures, and I'll tell you a story about Ramsey numbers. For now I just want to point out, as you might imagine, that the Ramsey numbers are pretty hard to compute exactly. Mostly, we have to settle for inequalities. In fact, the only non-trivial Ramsey number known involving more than 2 colors is R(3, 3, 3), and it is, in fact, equal to 17. Let's content ourselves with showing that R(3, 3, 3) is at most 17. This will help you to understand what R(3, 3, 3) is. We'll do this by proving that if we 3-color the edges of a K_{17} in any way we want, we're guaranteed to get a monochrome triangle. At this point, you're probably nervous and confused and thinking, "How am I going to visualize a K_{17}?" but I think you'll be delighted by how easy the argument becomes.

What we're really doing, and this is a fundamental idea, is we're taking a very large, complicated thing like a K_{17}, and we're viewing it as a superstructure with something smaller that we already know. We're going to take that K_{17}, and we're going to extract out of it a K_6 because we know things about K_6's. Here's how it works. We'll work exactly the way we started before. Start with a K_{17}. Visualize 17 dots, join all the vertices, and color those edges in 3 different colors, red, white, and blue. Then let's contemplate one vertex, which we'll call v. By the pigeonhole principle, the intermediate pigeonhole principle, if you look at that vertex v, it has 16 edges emanating from it, and there are 3 colors; those are our 3 categories. If we divide 16 by 3, we get 5 and a fraction, and when we take the ceiling, it gives us 6. At least 6 of those 16 edges emanating from that vertex must be the same color, without loss of generality, red. Now, we look at the vertices that are the endpoints of those red edges. There are 6 of them, along with v. We actually have 7 vertices in total. If you look at those 6 vertices, a, b, c, d, e, f, just ask yourself: What colors could you join among those 6 vertices? You have lots of choices. You could do red, but as soon as you did red, you'd be done. What if you did white or blue? Then you would have the vertices a, b, c, d, e, and f only colored in the 2 colors, white and blue. That reduces us precisely to the original problem of 6 vertices, and we know that if we have 6 vertices and we color the edges in 2 colors, we're guaranteed a monochrome triangle. In other words, either there will be a red triangle, or if there aren't any red triangles, what we're left with will turn out to be a K_6 Ramsey problem, which we've already solved. We were able to force a monochrome triangle starting with those 17 points.

Why did all this stuff work? Why was it useful? We were able to use the pigeonhole principle to solve strange problems, and you'll notice, some of those number theory problems we solved were, in many ways, kind of contrived. What's the unifying idea behind these pigeonhole problems? The unifying idea is that we try to strive for commonality, for coincidence, for uniformity, for equality. These are often things that we want to find in a problem. We're trying to get 2 things that, perhaps, are equal or maybe 2 things that are balanced in some way. For example, what makes something congruent to 0 (mod 17)? The way you get a single thing congruent to 0 (mod 17), you could look at 2 things that are equal (mod 17) and then subtract them. A logical method that forces 2 things to be in the same category, what it does is it gives you a powerful penultimate step to solve problems of this kind. That, in a sense, is what makes the pigeonhole principle useful. It's a powerful way of solving problems for which the penultimate step involves putting 2 things in the same category.

In the next lecture, we're going to move to, in my opinion, a far more profound tactic. What this tactic will do is rather than give us a way to put 2 things into the same category, it gives us a way to look at a single thing, a single entity, and ask the question: What about this entity is staying the same, and why is that important for problem solving? We'll explore this very important issue in the next lecture when we look at the tactic of invariance.

Lecture Fifteen

The Greatest Unifier of All—Invariants

Scope:

Invariants are central to mathematics, yet most laypeople have never heard of them. In this lecture, we show how the concept of invariants contains both symmetry and parity. We tweak it to look at monovariants and use these to study some interesting games. We also continue our study of modular arithmetic, which we now see is merely a special case of the grand unifying principle of invariants.

Outline

I. An invariant is any quantity or quality that stays unchanged.

II. A geometric example is the power of a point theorem.

- **A.** In any circle, when 2 chords intersect inside a circle, they obey the equation $AE \times EB = DE \times EC$.
- **B.** Likewise, when chords intersect outside the circle, $EA \times EB = ED \times EC$.
- **C.** These 2 theorems seem like related results about intersecting lines and circles, but in fact they are actually manifestations of a single fact.
  1. For any fixed point P and any fixed circle, draw any line through P that intersects the circle in points X and Y. Define the power of P to be the quantity $(PX)(PY)$.
  2. The power of a point theorem says that for any fixed circle and any fixed point, this quantity is invariant, no matter which line we choose.
  3. This invariant formulation is true no matter where the point is located.

III. Here is a classic brainteaser called the hotel room paradox that exploits our native instinct to look for invariants. Three women check into a hotel room that advertises a rate of $117 per night. They each give $40 to the porter, and they ask him to bring back $3. The porter goes to the desk, where he learns that the room is actually only $115 per night. He gives $115 to the desk clerk and gives the guests back each $1, deciding not to tell them about the actual rate. Thus the porter has pocketed $2, while each guest has spent $39, for a total of $2 + (3 \times 39) = \$119$. What happened to the other dollar?

A. The question is not what happened to the dollar, but how do the variables relate to one another? What is invariant?

B. The money the guests paid is equal to the amount that the hotel received ("hotel" means the porter and the desk). In other words, if g, p, and d are respectively equal to what the guests pay, what the porter pockets, and what the desk receives, then the quantity $g - p - d$ is an invariant, always equal to 0.

C. The "paradox" is the fact that $g + p = 119$, which is close to 120. But this quantity is not invariant, and it can assume many values. For example, if the actual price of the hotel was $100, then the porter would give the desk $100, return $3 to the guests, and keep $17. Then $g + p = 119 + 17 = 126$, which seems less paradoxical.

D. What confuses people is that they think $120 is invariant. And it is, as long as you think clearly about it: 120 is not the invariant amount of dollars "in circulation." Instead, 120 is the invariant "net worth" of the guests: the sum of the dollars they possess, the value of their room, and the amount that the porter stole. These numbers indeed add up to 120.

IV. Here is another classic puzzler: Bottle *A* contains a quart of milk, and bottle *B* contains a quart of black coffee. Pour a half-pint of coffee from *B* into *A*, mix well, and then pour a half-pint of this mixture back into bottle *B*. What is the relationship between the fraction of coffee in *A* and the fraction of milk in *B*?

A. It is possible to do this with algebra, and when we do so, we discover the surprising fact that the fraction of coffee in *A* is equal to the fraction of milk in *B*.

B. But this can made obvious once we think about invariants. The coffee and milk both satisfy conservation of mass (really, volume). So if bottle A has x ounces of coffee "pollution," then bottle B is missing x ounces of coffee and thus has x ounces of milk pollution. Both bottles are equally polluted.

V. Here is an example using a parity invariant: Let $a_1, a_2, \ldots, a_n$ represent an arbitrary arrangement of the numbers 1, 2, 3, … , n. Prove that if n is odd, the product $(a_1 - 1)(a_2 - 2)\cdots(a_n - n)$ is even.

A. First, we give a solution that uses a pigeonhole argument.

B. An alternative solution uses invariants. Given any permutation $a_1, a_2, \ldots, a_n$, observe that the quantity $(a_1 - 1) + (a_2 - 2) + \cdots + (a_n - n)$ is invariant; namely, equal to zero!

C. Thus the terms in the product we are interested in add up to zero, and there are an odd number of them. Clearly, they cannot all be odd, since a sum of an odd number of odd numbers is always odd, and zero is even. So one term must be even.

VI. Here is an example that uses congruence: At first, a room is empty. Each minute, either 1 person enters or 2 people leave. After exactly an hour, could the room contain 100 people?

A. Get hands dirty to work out examples.

B. Is there anything that all the possible outcomes have in common at a fixed time?

C. Yes! Suppose the population is p at some time. A minute later, the population will be either $p + 1$ or $p - 2$. Notice that these numbers differ by 3. This pattern will continue indefinitely, so at any fixed time, there will be many different outcomes, but they will all be congruent (mod 3). In other words, for any fixed time, population is invariant modulo 3.

D. One possible outcome is a population of 60. Thus, all possible outcomes in 60 minutes are congruent to 60 (mod 3). Since 60 is a multiple of 3, and 100 is not, we conclude that the population cannot equal 100 after an hour.

VII. A monovariant is a quantity that changes, but only in one direction. Monovariants are useful for studying evolving systems.

A. Here is a simple example (due to John Conway), called Belgian waffles. Two people take turns cutting up a waffle that is 6 squares × 8 squares. They are allowed to cut the waffle only along a division between the squares, and cuts can be only straight lines. The last player who can cut the waffle wins. Is there a winning strategy for the first or second player?

1. This is a fake game. There is no strategy because of this simple monovariant: Each move increases the number of pieces by 1!
2. You start with 1 piece (the whole waffle), so the game ends in 47 moves no matter what the players do!

B. The next problem illustrates the power of monovariants to cut through the complexity of evolving systems. At time $t = 0$ minutes, a virus is placed into a colony of 2009 bacteria. Every minute, each virus destroys 1 bacterium, after which all the bacteria and viruses divide in 2. For example, at $t = 1$, there will be $2008 \times 2 = 4016$ bacteria and 2 viruses. Will the bacteria be driven to extinction? If so, when will this happen?

1. There are complicated algebraic formulas that will do the trick, but monovariants are a better way.
2. Let b and v be the respective populations at a certain time, and let b' and v' be the populations 1 minute later.
3. It is easy to see that $b' = 2(b - v) = 2b - 2v$ and that $v' = 2v$. The trick is to manipulate these quantities to recover something that is almost constant.
4. If we divide, we get $\frac{b'}{v'} = \frac{2b - 2v}{2v} = \frac{b}{v} - 1$.
5. In other words, the ratio b/v is a monovariant: It decreases by 1 each minute.
6. At $t = 0$, the ratio is $2009/1 = 2009$. In 2009 minutes, it will decrease to 0, and the bacteria will be wiped out.

Suggested Reading:

Engel, *Problem-Solving Strategies*, chap. 1.

Zeitz, *The Art and Craft of Problem Solving*, sec. 3.4.

Problems:

1. A graph that can be drawn so that edges do not cross is called a planar graph. For example, a typical picture of a K_4 has the 2 diagonal edges crossing, but it is possible to draw this graph where one diagonal is "inside" and the other is "outside," so that K_4 is planar. Given any planar graph, it is easy to count the number of vertices (v), edges (e), and regions bounded by edges (r). Discover an invariant involving these variables.
2. Can you find distinct integers a, b, and c such that $a - b$ evenly divides $b - c$, $b - c$ evenly divides $c - a$, and $c - a$ evenly divides $a - b$?

Lecture Fifteen—Transcript

The Greatest Unifier of All—Invariants

In this lecture, we're going to look at invariance. Simply put, an invariant is any quantity or quality that stays unchanged. We've seen invariants before. For example, in the frog problem, with the jumping frogs, in this one, the parity of the coordinates of each frog was invariant. Symmetry is another example. If an object has reflectional symmetry, it is invariant with respect to reflection. You can also talk about invariance in a physical way. There's conservation of mass; mass is invariant. Conservation of angular momentum—you can use physical intuition with invariance. In other words, invariance is a very big topic. In my feeling, it's the mother of all tactics because, after all, symmetry is just a special case of invariance. In fact, if I had to pick the single most important word in all of mathematics, it would be the word invariance. You might wonder why you haven't heard about invariance in your education. I'm oversimplifying here, but that's because most mathematical education traditionally is exercise dependent, but the central idea of this course is the analytic approach to problem solving, looking at things from a higher level. We want, as often as possible, to have a bird's-eye view. Invariants are a very, very high-level way at looking at many, many problems. It's important to cultivate this attitude of deconstructing and analyzing problems that are solved and even [those that are] unsolved to see the central underlying ideas behind them. In this lecture, we'll see several examples of invariance in action, and we'll also look at a related concept, that of the monovariant, which is like an invariant. But it doesn't stay unchanging; it just changes in only one direction, hence the name monovariant. The problems that we'll look at will be mostly pretty simple compared to some of the more elaborate ones we've seen so far, but don't worry; we'll ramp up the complexity as we apply these concepts using invariants in later lectures.

For now, let's start with a simple geometric example that you might remember from school. It's a theorem called the power of a point, although, you probably didn't hear that theorem name. Here's the first example. If you look at a circle and the cords are intersecting, say, at that point E, you probably learned that the product $AE \times EB = EC \times ED$. In other words, the product of the lengths AE and EB equals the product of the lengths CE and ED. You might also know another geometry theorem, where you have a point outside the circle, and in this case, it would say $EA \times EB = ED \times EC$. You might not remember them; you might sort of say, "Oh yeah, I vaguely remember learning that in geometry." They look like similar theorems that are related results about lines intersecting circles, and if you learned them in your geometry class, I'm sure that's how it was taught. Here are 2 theorems about lines intersecting circles, but in fact, these 2 things are actually manifestations of a single fact, a single theorem. Let's define something called the power of a point. For any fixed point P and any fixed circle, let's draw any line through P that intersects the circle, say, in points X and Y. What we will do is define a quantity called the power of P, and it's equal to the quantity $(PX)(PY)$. It's the product of the length of those 2 line segments. You take a point, look at the line that hits the circle, find the lengths to each end of the circle, and multiply those 2 lengths. That's the power of the point. What the power of a point theorem says is that for any fixed circle and for any fixed point, this quantity is invariant no matter which line we choose. This invariant formulation is true no matter where the point is located. If it lies on the circle, then the power of the point will be equal to 0, and the theorem has no content. If the point is inside the circle, it takes the first form that we saw. If the point if outside the circle, we automatically get the second theorem. The main point here is not to review geometry but just to focus on the fact that an invariant formulation is a very powerful and very compact way to get to the heart of what's really going on in a problem.

Let's change gears to an old brain teaser. This is a problem that you may have heard before. What it is exploiting is one's native instinct to look for and hope for the existence of invariants. Unfortunately, there are many ways to get bamboozled because there are plenty of false invariants. Here's the situation: 3 women check into a hotel room that advertises a rate of \$117 per night. They each give \$40 to the porter and ask him to bring back \$3. The porter returns to the desk, where he finds out that the room is actually only \$115 per night. He gives \$115 to the hotel desk clerk, and then he returns to the room and gives the guests back each \$1, deciding not to tell them about the actual rate. Thus, the porter has pocketed a \$2 tip, while each guest has spent $40 - 1$, or \$39, which is a total of \$2—the porter's—plus 3×39, which is \$117, and that adds up to \$119. The question is: What happened to the other dollar?

This classic old puzzle is a great example of how the invariant point of view really clarifies issues. The question is not what happened to the dollar, but how do the variables actually relate to one another? What is the actual invariant? The money the guests paid is equal to the amount that the hotel receives. The hotel, in this case, means the porter plus the desk. In other words, if we let g, p, and d, respectively, equal what the guests pay, what the porter pockets, and what the desk receives, then the quantity $g - p - d$ is an invariant and always equal to 0. The paradox is the fact that $g + p$ in our problem equals 119, and that's really close to 120. But so what? The quantity $g + p$ is not invariant, and it can assume many, many values. For example, if the actual price of the hotel was $100, then the porter would give the desk $100, return $3 to the guests, and keep $17. Then $g + p$ would equal $119 + 17$, which is 126, and that doesn't really seem like much of a paradox. What confuses people is that they think 120 is some kind of invariant, and it actually is as long as you think clearly about it. [Note,] 120 is not the invariant amount of dollars in circulation. Instead, 120 is the invariant net worth of the guests. It's the sum of the dollars they possess plus the value of their room and the value of the service rendered, in other words, the involuntary tip that the porter stole. These numbers are, respectively, 3, 2, and 115, and indeed, they add up to 120. If you think about it, this hotel room paradox is just a clever repackaging of the old finger-counting trick that my grandpa used to do to me. Maybe your grandpa did it, too, where you go 1, 2, 3, 4, 5, 6, 7, 8, 9, 10 with a little child, and they say, "Yeah, I have 10 fingers." Then grandpa says, 10, 9, 8, 7, 6 plus 5 is 11. Now you have 11. It might make a child cry if they're very mathematically sensitive.

Here's another classic problem. It's about mixing drinks. Imagine bottle *A* contains a quart of milk and bottle *B* contains a quart of black coffee. Pour a half-pint of coffee from *B* into *A*. Mix it well, and then pour a half-pint of this mixture back into bottle *B*. Now both bottles have a quart of fluid. The question is: What's the relationship between the fraction of coffee in *A* and the fraction of milk in *B*? It's a fun question to ask people. Will the one that started with coffee have more milk, or will the one that started with milk have more coffee? You get lots of different guesses from people. You should try to think about this on your own. The analysis isn't hard. You could just do this with a careful calculation or, in the general case, where the half-pint could be any volume, with some algebra. When we do so, we'll discover a surprising fact that the fraction of coffee in A is equal to the fraction of milk in B. If you start, for example, with 1 unit of coffee and pour 1/8 into milk, then the milk will be 1/9 coffee and 8/9 milk. It will have proportions 8 to 9 milk to coffee. Then, if we pour 1/8 of a unit of that mostly milk mixture back into the coffee, then the actual amount of milk going back to the coffee is 8/9, the fraction of milkiness, of 1/8, the actual amount of fluid. When you do the math, that's $8/9 \times 1/8$, which is 1/9. The coffee is now 1/9 milk and 8/9 coffee. It's just the reverse of the other bottle. That analysis should confuse you. It's fractions, and it's hard to keep track of what's going on.

Let's make it obvious. We can make it obvious when we think about invariants. The coffee and milk both satisfy conservation of mass or, literally, volume. If bottle A has x ounces of coffee pollution in it, then bottle B has to be missing x ounces of coffee and, thus, has x ounces of milk pollution displacing that. Both bottles, therefore, are equally polluted. That still might not be obvious because it wasn't obvious to me. I had to get my much smarter friend, Doug, to help me to truly see what was going on when he said, "You know what you should do, Paul? Just imagine that instead of fluids, one bottle is filled with white ping-pong balls, and the other bottle is filled with black ping-pong balls." That really helps one to grasp the notion of conservation of mass. When I first heard about those ping-pong balls, for me, it was an epiphany. Suddenly, I understood this problem even though I could do it algebraically. Remember, I've said before algebra isn't such a great thing. It often allows you to figure out how to do something, but you don't know why you're doing what you're doing. Ping-pong balls teach you why.

Invariants can be found in almost any field of math and physics, as well. You can think about parity invariants, congruence invariants, coloring invariants, and physical invariants. Why don't we look at a parity example? This one was one of the first of the Hungarian problems at the beginning of the 1900s, and it's a great example of a multifaceted problem, a problem that's not that hard and a problem that can be solved in many, many different ways. At this point, you've learned a lot of problem-solving strategies and tactics, and you might wonder: What do I use when? Some problems are very, very specific. They really are only, say, pigeonhole problems, but there are many problems, as well, where many different tactics and many different strategies can solve the problem. It's good to keep in mind that you might not know how to begin attacking a problem, but you now have an arsenal of ideas and you should just try and see what works. Let's look at the problem. Imagine you have a sequence of n arbitrary integers, the numbers from 1 to n. We take the integers from 1 to n, and we make an arbitrary rearrangement of them. Let a_1, a_2, a_3, down to a_n be just a rearrangement of the numbers from 1 to n, a permutation, in other words. You take the numbers 1, 2, 3 up to n, and we just rearrange them in some way. We call the first such number a_1, the second number a_2, and so on. What we want to prove is that if n is odd, the product $(a_1 - 1)(a_2 - 2)(a_3 - 3)$ and so on down to $(a_n - n)$ is an even number. You see the word even, you know to think about parity.

Let's do an example. Let's say $n = 5$, and let's write our permutation of the numbers from 1 to 5 as 4, 3, 1, 5, 2. The appropriate product will be $(4 - 1)(3 - 2)(1 - 3)(5 - 4)(2 - 3)$, and it's easy to do that computation. It's easy to see that it's even, and we don't even need to do the full computation because it has one even factor in there; $1 - 3$ is -2. In other words, the penultimate step for a product being even is for at least one of the factors of the product to be even. This is a single get your hands dirty example, but it showed you really what's going on. We just need to show that at least one of the numbers in the product must be even, and then we'll be done. We can do this by counting odds and evens and noticing that because n is odd, more than half of the a_n are odd. For example, if n is 17, if you think about it, the numbers 1, 2, 3, 4, 5 up to 17, 9 of them are going to be odd and 8 of them are going to be even. That's what happens with odd numbers. They are overabundantly odd in this way. Thus, we will fail at pairing a_k with k in such a way that the 2 numbers will always have different parity. There will always be at least one pairing where a_k and k are both odd, forcing $a_k - k$ to be even and, hence, the entire product to be even. This parity matching arrangement can be made clearer if you just imagine that the odd terms are colored red; if you permute them, and then what you want to try to do is write the numbers 1 through n under them in such a way that the reds are going to not match up. There will always have to be at least one red-red match-up. We can't avoid a red-red match-up because there are just too many reds. In other words, we're doing sort of a pigeonhole principle argument using parity.

Those are good ways of looking at the problem. We found a simple parity analysis using just the tiniest of a pigeonhole argument. It's a great solution; there's nothing wrong with it. But here's another way to look at it, which in some ways is just wonderful because it's unexpected and it teaches you a different way to think of a problem, to think of it from a higher point of view. It's inspired by thinking about telescoping sums. Let's consider that permutation $a_1, a_2, \ldots a_n$, and look at those differences $(a_1 - 1)$, $(a_2 - 2)$, and so on, and add them up. In other words, just look at the quantity $(a_1 - 1) + (a_2 - 2) + (a_3 - 3)$ all the way up to $(a_n - n)$. If you get rid of the parentheses, you're adding up all the a_1 through a_n, and you are adding up all the 1 through n and subtracting that from it, but the a_1 through a_n are just the numbers from 1 to n in some permutations, so when you add them up, you'll get the same sum. You're going to be adding up a bunch of things and subtracting the same numbers. In other words, that quantity, no matter what the permutation is, is an invariant, namely, the invariant 0. "So what?" you might say. But 0 is even, and how many terms are in the sum? An odd number of terms. If you have an odd number of terms adding up to 0, one of them must be even because if they were all odd, then you would be adding an odd number of odd numbers and that would be odd, a completely different way of solving the problem. It still used parity, but it used an invariant as your entry point.

Let's take a look now at a congruence example. This is a problem involving sort of an evolving structure. At first, a room is empty. Each minute, either 1 person enters or 2 people leave. Randomly we choose this. After exactly an hour, could the room contain 50 people? A lot of my problem-solving colleagues, if they looked at a problem like that, they'd say, "That looks like a Russian-style problem." There is something Russian about it, and if you've been around the block with these problems enough, you say, "That looks like some sort of a problem involving an invariant." We want to see what's invariant and will it agree with what's happening? Perhaps it's a monovariant. Let's analyze it. Clearly, the population can be all sorts of numbers as it evolves. Remember, at each minute, either 1 person enters or 2 people leave. Here are a few examples. One case you could just have 1 person enter each minute, so we go 1, 2, 3, 4, 5, 6, 7, 8, 9, but we could also go 1 person, then 2, then 3, then 4, then 2 people leave, bringing it down to 2, then 1 person enters, bringing it up to 3, then 4, then 5, then 6. Or in the third case, we could go 1, 2, 3, 4, 2, and then if 2, leave down to 0, and then 1, 2, 3 again and so on. If you go out far enough, you can have many, many different sorts of contingences. This is very easy; [we're] getting hands dirty at this point. What's going on? Is there anything that you can notice about the possible outcomes? Is there anything that they have in common at any fixed time? Yes. Suppose the population is p at some particular time. A minute later, the population will either become $p + 1$, adding 1 person, or $p - 2$, if 2 people leave. Notice that these 2 numbers differ by 3. This pattern will continue indefinitely; at any fixed time t, there will be plenty of different outcomes, but they will all be congruent modulo 3. That's what it means to differ by 3 or by multiples of 3. In other words, for any fixed time, the population is invariant modulo 3, and in an hour's time, one possible evolution is for the population to just add 1 each minute for a total of 60. Thus, all possible outcomes in 60 minutes must be congruent to 60 modulo 3, but since 60 is a multiple of 3 and 50 is not a multiple of 3, we conclude that the population could never equal 50 in exactly one hour. That's kind of a strange example, but here's a stranger one.

Instead of looking at a congruence invariant, let's look at something more metaphorical, sort of a structural or systemic invariant. Do you remember the bug problem, when we talked about symmetry of the 4 bugs that crashed into each other? We figured out how long it took for them to crash, but now let's ask ourselves: What's the total rotation before they collide? Imagine that you're one bug, and you're facing east. As you start to move, you might eventually face north, then you've turned 90 degrees counterclockwise. If you kept turning some more, maybe you would turn another 90 degrees. How far do you turn? Imagine you're a bug; you're keeping track of your turning until you crash into the other bugs. How much rotation would that be? If you look at the picture, there's no way of telling. The picture just isn't detailed enough. Let's imagine the starting position, and what you really have at any given time is: Bugs are always at the vertices of a square. The square is shrinking, but the bugs are always at the vertices of a square until that square just crashes and becomes a single point. Imagine that you're a bug, and you just travel until you've rotated 90 degrees. After that's happened, what's your configuration? You still are one of 4 bugs that are along a square. You're still 4 vertices of a square, and from your point of view—pretend you have no sense of scale—what's going to happen? If you turn 90 degrees from your initial square, now you're at some square. You're going to turn another 90 degrees, and then where will you be? You'll all be at some square. Then what are you going to do? You'll turn another 90 degrees. How many of these 90 degrees will you turn? No one can stop you from just turning 90 degrees indefinitely. In other words, it takes a certain amount of time to rotate 90 degrees, but as far as the bugs are concerned, you're back to the starting point. You're just going to keep on rotating 90 degrees forever. The time it will take will get shorter and shorter, but the geometric configuration is invariant. The size is changing, but the shape is not, and in terms of angles, there's always going to be an opportunity to rotate more. We get something that's paradoxical almost. It will take a finite amount of time for the bugs to crash, but within that time, they'll be spinning faster, and faster, and faster, and their total rotation will exceed any finite number, in other words, will be infinite.

Now let's turn to the concept of monovariants, which as I said before, is closely related to invariants. A monovariant is a merely a quantity that changes but only in one direction. Monovariants are very useful for studying evolving systems and proving that they terminate or that they have certain states that are impossible. A classic example of this is the famous checker problem of John Conway, who you've heard about, and we will discuss this in depth in a later lecture. First, let's look at a much simpler example, and this is also due to Conway. It's called Belgian Waffles. Imagine a game where 2 people take turns cutting up a waffle that is 6×8 squares in size. It's like one of those combinatorial games we've played before. You're allowed to cut the waffle only along a division between the squares, and your cut can only be a straight line. The last player who can cut a piece of waffle will win and get to eat all the pieces. Is there a winning strategy for the first or for the second player? We know how to analyze this with oases and deserts if we wanted to, but let's just look at a couple of simple games. Here's one: We take a cut down the vertical line. Then we cut the left piece down a horizontal line, and then we cut the top left piece all the way at the left. That's a sample of a few moves of this game. The game is fun to analyze, but if you try to take it too seriously, you get snowballed. Because this is an example of a fake game. There is no strategy because there's a simple monovariant involved. Each move that occurs increases the number of pieces on the board by 1. You start with 1 piece, the whole waffle, and then you make a move and there are going to be 2 waffle pieces. You make another move, and there will be 3 waffle pieces. It doesn't matter who makes what move. Each time a move is made, you'll have 1 more piece. The number of pieces will keep increasing, and the game will end in 47 moves, when we have a total of 48 pieces, no matter what the players do. The first player will win because 47 is odd. This game is a silly game, and when I first learned about it at one of the Math Olympiad training programs, I was told that the smarter you were, the harder it was for you to figure out this game. Sure enough, I was the first person to figure out how the game worked. I didn't feel triumph because I was already warned that if I figured it out, it would just mean that I was dumb.

Let's look at one more monovariant. This problem allows you to work with a much more complicated evolving system, where there could be potentially hard computational or algebraic problems, but taking a monovariant point of view allows you to just sort of cut through the complexity. It involves viruses and bacteria. I call it a Darwinian struggle. At time t = 0 minutes, a virus is placed into a colony of 2,009 bacteria. Every minute, each virus destroys one bacterium apiece, after which, all the bacteria and all the viruses that are alive divide in 2. For example, at t = 1, there will be 2008 × 2, or 4016, bacteria, because 1 bacteria was eaten, but then they'll double, and there will be 2 viruses. At time t = 2, we'll subtract 2 from the 4016 bacteria, so there will be 4014. Multiply that by 2. That will be the amount of bacteria, and then there will be 4 viruses, etcetera. The question is: Will the bacteria be driven to extinction, or will the bacteria swamp the viruses? What's going to happen? It certainly looks like the bacteria are doing pretty well. We have over 4000 of them after just a few periods of time. You could do lots of complicated algebraic formulas involving geometric sums, but let's not do that. Algebra is too hard. We'll need to use a little bit of algebra, but not a lot. Let's let b and v be the respective populations at a certain time of bacteria and viruses, and let's let b' and v' be the populations one unit later, 1 minute later. It's easy to see that $b' = 2(b - v)$ because you have to subtract the number of viruses from the bacteria because each virus eats a bacterium. So $b - v$ is the new amount of bacteria, but then we double it, so $2(b - v) = 2b - 2v$. Meanwhile, v' is easy; it's just equal to $2v$ because every generation, the viruses just double in size. The trick is to manipulate these quantities in such a way as to recover something that is either constant—in which case, we'd get an invariant—or something that's almost constant, and then we could hope for a monovariant. We just try some simple things. How about dividing b' by v'? If we do that, then we get algebraically the quantity $(2b - 2v)/2v$, and then if you just look at the fractions, you'll get $2b/2v - 2v/2v$. In other words, when you cancel, you'll get $b/v - 1$. That's beautiful because what it tells you is that the ratio of the bacteria to viruses at 1 minute later is equal to the ratio at the starting point minus 1. In other words, the ratio b/v is a monovariant; it decreases by 1 each minute, and now the problem is solved. At time $t = 0$, the ratio is 2009 to 1, which is 2009. We know it will take exactly 2009 minutes for that ratio to decrease to 0, and then the bacteria will be wiped out. Of course, in the next generation, the viruses will all starve, so the bacteria do get their revenge. The real point here is how easy it was to look at this problem with a monovariant.

These examples, for the most part, were fairly simple in terms of algebra or the length of solution, but they varied over many mathematical topics. The point is that the idea of invariance is a tremendous unifier. You can use it almost anywhere. It's a unifier that takes the idea of things such as the natural point of view, or symmetry, or parity to a much higher level of abstraction. We'll see this in action as we apply invariants and monovariants in much more elaborate ways in later lectures.

Lecture Sixteen

Squarer Is Better—Optimizing 3s and 2s

Scope:

In this lecture, we return once again to our great friend symmetry to explore questions of distribution and optimization. Along the way, we develop a new proof method, algorithmic proof, where we imagine a sequence of steps guaranteed to solve our problem. Our anchor problem is an International Mathematical Olympiad problem about a maximal product.

Outline

I. Our anchor problem: Determine, with proof, the largest number that is the product of positive integers whose sum is 1976.

  A. Intuition may tell us to try a square: $988 \times 988 = 976{,}144$. But we can do better. For example, $987 \times 987 \times 2 = 1{,}948{,}338$.

  B. Clearly we need more investigation! First, let's consider simpler, more constrained questions.

II. Warm-up: A rectangle is made of 12 inches of wire. What should the dimensions be to maximize the area?

  A. One solution is to appeal to symmetry: Obviously, the rectangle of largest area is the most symmetrical. So the dimensions are 3×3.

  B. How do we do this rigorously? General question: If x and y have fixed sum S, what is the maximum value of the product $P = xy$, and what will x and y be when this maximum is attained?

  C. Conjecture: Maximum is $(S/2)^2$, when $x = y = S/2$.

  D. This can be proven with calculus.

III. But a better way to prove this squarer-is-better principle is with a picture.

  A. Let $S = x + y$. In other words, S is the diameter of the semicircle.

  B. By similar triangles, $g/x = y/g$, so $g = \sqrt{xy}$.

- **C.** The maximum value of xy is attained when $x = y = S/2$.
- **D.** As the distance between 2 positive numbers decreases, their product increases, provided that their sum stays constant.
- **E.** Note that this is a dynamic principle.

IV. A reformulation of this is the 2-dimensional arithmetic-geometric mean inequality (AM-GM): If x and y are nonnegative, then $\frac{x+y}{2} \geq \sqrt{xy}$, with equality attained when $x = y$.

- **A.** The AM-GM is also true in higher dimensions. For example, in 3 dimensions, the statement is $\frac{x+y+z}{3} \geq \sqrt[3]{xyz}$, which is very hard to prove using algebra alone.
- **B.** The 2-dimensional AM-GM is not too hard to prove with algebra. It is equivalent to $(x-y)^2 \geq 0$, which is certainly always true.
- **C.** But it is hopeless to prove the general n-dimensional AM-GM with algebraic methods.

V. However, we can use the 2-dimensional squarer-is-better principle to prove AM-GM in any dimension if we leave algebra behind and instead view the problem in terms of physics.

- **A.** The original formulation for n variables is as follows: Let $x_1, x_2, \ldots, x_n$ be positive real numbers. Then $\frac{x_1 + x_2 + \cdots + x_n}{n} \geq \sqrt[n]{x_1 x_2 \cdots x_n}$, with equality only when the numbers are equal.
- **B.** Reformulation: Let the sum S of $x_1, x_2, \ldots, x_n$ be fixed, and let the product be P. Then the AM-GM asserts that $S/n \geq \sqrt[n]{P}$, with equality only when the x_i are all equal. Raising by the nth power, we get $P \leq (S/n)^n$.

VI. We prove this reformulated version by using an algorithmic method with weights. We make the process of finding an optimal product a physical process. The breakthrough idea is to make the x_i unit weights placed on a number line. Then their average value corresponds to their balancing point!

VII. Let's return to the 1976 International Mathematical Olympiad problem. We have seen how powerful symmetry is. Our instinct that equal values optimize products has been rigorously proven. But what if we are unable to make the parts equal?

- **A.** There are 2 difficulties: We are not told how many numbers are in the product. And even if we were, we could not guarantee that the values would be integers if we made them all equal.
- **B.** Begin your investigation by replacing 1976 with smaller values.
- **C.** Conjecture: Use only 3s and 2s.
- **D.** Why? Once again, try an algorithmic method. Assume that we have a maximal product of integers with a fixed sum and that we have numbers that are not 3s and 2s.
- **E.** Conclusion: The only numbers possible in an optimal product are 2s and 3s.
  - **1.** Next, notice that if you have three 2s, you can replace them with two 3s.
  - **2.** So that gives us the optimal breakdown: all 3s, unless you have to have some 2s, but never more than two 2s. For example, 12 breaks into $3 \times 3 \times 3 \times 3$, but 13 gives us $3 \times 3 \times 3 \times 2 \times 2$.
- **F.** And 1976 breaks into $3^{658} \times 2$.

Suggested Reading:

Kazarinoff, *Geometric Inequalities*.

Niven, *Maxima and Minima without Calculus*.

Problems:

1. If you have a fixed length of wire and you are to make a rectangle of maximum area, you know that a square is optimal. But what if one side of your rectangle is already provided? For example, suppose you are building a rectangular fence, one of whose sides is a river? What will the optimal dimensions be? (Hint: symmetry.)
2. Use the arithmetic-geometric mean inequality and symmetry to prove the nice inequality $(a + b)(b + c)(c + a) \geq 8abc$, for positive a, b, and c.

Lecture Sixteen—Transcript

Squarer Is Better—Optimizing 3s and 2s

In this lecture, we will return to our old friend symmetry to explore questions of distribution and optimization. Along the way, we will develop a new proof method called algorithmic proof, where we imagine a sequence of steps, an algorithm, which [is] guaranteed to solve our problem. Essentially, this is proof by writing a computer program. Our anchor problem is from the IMO. I'd like you to guess which year it's from. Here's the question: Determine, with proof, the largest number which is the product of positive integers whose sum is 1976, in other words, 1976. Now, your intuition says, perhaps try a square. We'd like to find 2 numbers that add up to 1976, and if they're equal, it's sort of like finding the area of the square, so we could go 988×988, and as you know, that's 976,144, and that's pretty large. What if you did more than 2 parts? After all, the problem didn't say how many parts to break 1976 into. What if we had, say, a 987 and a 987? That leaves 2 left over, and now, if you look at the product, we have $987 \times 987 \times 2$, which is 1,948,338. That's a lot bigger. It's twice as big. What if we broke it into 4 pieces? Say 658, 658, 658, and 2. Then we'd get 569,780,624. The problem certainly needs more investigation, but before we do that, let's consider simpler, more constrained questions.

Here's a very typical problem that you might see in a calculus class. A rectangle is made of 12 inches of wire. What should the dimensions be to maximize the area? One solution, which in my opinion is the very, very worst one, uses calculus. Another solution, which is better, but it's not rigorous, is just to appeal to symmetry and to say, obviously, the rectangle of largest area would have to be symmetrical, so it would have to be a square. The dimensions are 3×3, and we say obviously. Appealing to symmetry, it's very laudable—I have nothing against symmetry, as you know—but appealing to it is sort of pathetic. We're using the more or less correct assumption that good things happen when stuff is symmetric, but that's not rigorous. It's good to keep in mind that, yes, good things happen when things are symmetric, but that's not a proof. With my college students, I'll often ask them to give me different levels of rigor. I'll say, "Okay, if you just had to give a conjecture and you just had to get it right, that's one level." I might say, "Look, suppose you're at the gates of heaven and you'll be admitted if you can just answer a true/false question about a conjecture." But what if you won't get admitted unless you can prove rigorously the truth or falsity of your assertion? Those are very, very different levels of rigor. I'd like to move towards the latter level of rigor.

Here's the general question: If x and y have a fixed sum of S, and these are all positive numbers, what is the maximum value of the product xy? We'll call that product P, P for product [and] S for sum, and what will x and y be when this maximum is attained? The conjecture is the symmetric one. The maximum will be attained when x and y are just equal to half the sum, in other words, equal to the average value, and then the maximum would be that $S/2$, half the sum, times $S/2$. It would be $(S/2)^2$ or $S^2/4$. This is a very, very reasonable conjecture to make. I said before that proving this with calculus is the worst possible way. It's not an evil way to prove it, but it doesn't illuminate things. The typical calculus proof would go somewhat like this: Let x be one number, and $S - x$ is the other number. The product we wish to optimize is the algebraic quantity $x(S - x)$, and then there are calculus techniques for optimizing this. But you don't need calculus. We could just take that function and graph it. You could use a graphing calculator or free software on a computer, or you could buy fancy programs, such as Mathematica. If we just graph this function $x(S - x)$, and just look to see where it reaches a peak, we'll see that the symmetrical conjecture appears to be true. For example, here is a graph when $S = 12$, and you can see it goes from 0 all the way up to its height of 36, and then back down to 0, perfectly symmetrically. It's beautiful, and we're being rewarded by pondering symmetry just as we appealed earlier.

That's just an observation, and it's certainly a good one, but let's do something more interesting. Let's create a proof without words that uses geometry. It uses fairly simple plane geometry, in a beautiful context. Imagine a semicircle, and the diameter of the semicircle is the value S, which is some fixed value. We break it up into line segments of length x and y. So x and y add up to some fixed value S. Then, draw an altitude from where those 2 line segments meet. We look at that altitude, and we'll give it a length of g. If you look carefully at the triangles, you'll notice that triangle *abd* and triangle *bcd* are similar. The reason they're similar is because they are both right triangles. A triangle inscribed in a semicircle is a right triangle—that's an old theorem of geometry—and we dropped an altitude to create right triangles, as well. If we look at the proportions carefully, we get that $g/x = y/g$, and therefore, $xy = g^2$. This number g is called the geometric mean of x and y because g is equal to the square root of x and y, but it comes from this geometric process.

Let's think about g; g is equal to the square root of x and y, the square root of the product xy. If the product xy is maximal, the square root will be maximal, too. What we're really interested in finding out is: When is g at its largest? If you look at the picture, you can physically slide the point d, where x and y are joined, just slide it from left to right, and it will obviously reach its maximum right at the center of the circle when the length g is equal exactly to the radius of the circle and exactly to the average of x and y. Also, at that point, x and y will suddenly be equal. When x and y are both equal to $S/2$, that value g will reach its maximum, and therefore, the product will reach its maximum, as well. The idea is that as the distance between the 2 positive numbers decreases, as they get more and more the same, their product increases, provided that their sum stays constant. This agrees with our intuition. As a rectangle becomes more squarish, in other words, more symmetrical, it encloses area more efficiently.

We'll call this the squarer is better principle. It's a tactic, and it's a dynamic principle. It tells you that the product increases as you move values around, as the difference in absolute value between those values decreases. This inequality, that the average value of the 2 numbers, $(x + y)/2$, is always greater than or equal to that square root of xy, is called the arithmetic-geometric mean inequality, because $(x + y)/2$ is the average, the arithmetic mean, and the square root of xy is called the geometric mean. What we've proven with this picture with the semicircle is that that average value, $(x + y)/2$, the radius of the circle, is always greater than or equal to the square root of xy. This is called the 2-dimensional arithmetic-geometric mean inequality, or sometimes abbreviated as the 2-dimensional AM-GM. It's a well-known tool used by mathletes around the world, but often the mathletes just assume it's true for 3 dimensions or 4 dimensions. For example, if you were trying to imagine enclosing volume with a box with dimensions x, y, and z, where $x + y + z$ are constrained to have a fixed sum, you would imagine that the box of greatest volume would be one that is a cube where x and y and z are equal. Indeed, that's true, but what that means algebraically—if you convert this into algebra, you're trying to prove the inequality that $(x + y + z)/3$, the arithmetic mean, is greater than or equal to the cube root of xyz.

That's incredibly hard to prove algebraically because you'd have to cube both sides. Intuitively it makes sense, but a rigorous proof using algebra is quite difficult.

We could prove the 2-dimensional case using the algebra of polynomials without too much difficulty. For example, if you want to show that $(x + y)/2 \geq \sqrt{xy}$, just multiply by 2 and we have the equivalence assertion that $x + y \geq 2\sqrt{xy}$. If you square both sides, you get the equivalence expression that $(x + y)^2 \geq 4xy$, and then if you simplify the algebra, you'll get $x^2 - 2xy + y^2 \geq 0$, and that is equivalent to the square of $x - y$. In other words, $(x - y)(x - y) \geq 0$. That's always true because if you square any number, positive or negative, it'll be greater than or equal to 0, and it'll be equal to 0 if and only if $x = y$. But if you try to do that with 3 variables, you'll go nuts, and if you try to do it with 7 variables, you'll go nuttier. We need a different method.

What we're going to do is use this sort of physical process of sliding this altitude in a circle around this 2-dimensional squarer is better principle, and we'll try to ramp it up into higher dimensions. Our goal is to prove the arithmetic-geometric mean inequality in any number of dimensions. This is a very sophisticated inequality that we're about to demonstrate. What it says is: If you have n positive real numbers, say, x_1, x_2 up to x_n, their average, the sum divided by n, is greater that or equal to the geometric mean of those numbers, which is the nth root of their product. Here's a reformulation: Let's fix the sum, and we'll call it s. $x_1 + x_2 + x_3$ up to x_n is the fixed sum S, and let's look at the product P. Well, then the arithmetic-geometric mean inequality asserts that S/n, the average, is greater than or equal to $\sqrt[n]{P}$ with equality only when the values are equal. If we raise it to the nth power, we get the inequality that that average, $(S/n)^n$, is greater than or equal to P. In other words, the product P is less than or equal to the average raised to the nth power. In other words, the product is maximized whenever the x_i are equal, namely, equal to their average value, because the biggest the product could be is when you have n values at that average value. You have average times average times average, n times. That's the reformulation that we're going to use, but now we need to make another sort of recasting.

We're going to look at the problem in a physical way. We'll use our intuition, and we'll construct an algorithm for optimizing the product. What we'll do, our breakthrough idea, is that we'll make each of these x_1, x_2, x_3, and so on, we'll make them be weights, and we'll make them all be weights of the same value, say, 1 unit, 1 kilogram, perhaps. We'll place them on a number line based on their value. If x_1 is 3, we'll put it at the point $x = 3$ on the number line. If you think about that, we now have a sequence of values, maybe one at 3, one at 7, one at 1000. These are all equal weights, and if you imagine the number line as a ruler, then the average value is going to correspond to the balancing point of these weights. Here's an example: Suppose you have just the values 3 and 7. We put a unit weight at position 3 and another one at position 7. Clearly, the balancing point is right in the middle at 5, and that's the numerical average of 3 and 7. It's the same as 3 + 7 divided by 2. The physical version of our squarer is better principle is that if we slide those 2 weights together towards their balancing point, always keeping that same balancing point, the product will increase. In our case, it will go from 3×7 all the way to the maximum of 5×5. It will go all the way up from 21 up to 25.

Now let's try to use this to create what I call an algorithmic proof by contradiction. Let's assume that the sum of the n values—n could be a million—the sum of the n values is fixed. We want to prove the product is maximized when all the values are equal to their average, namely, S/n. The proof by contradiction would be: Suppose they're not all equal. Now we have something to work with. We have some weights on a number line, and they're not all in the same place; they're not all at the average. What we want to show is that they have to be at the average, but right now, we're assuming they're not all at the average, but we know their balancing point is the average, S/n. We'll call that number a. What we're going to show is that if the locations are not at the average and we can slide them in such a way as to increase the product, that's our proof by contradiction. In other words, we're assuming that we have a maximum product and they're not all at the average, but we'll slide some toward the average, make them more equal and increase the product, and then we'll be done. You can think of this as an algorithmic process of optimization. Just start out with some values, and we'll sort of create a recipe for optimizing their product, and it will end with them all being equal.

Here we go: If they're not all equal, then there are at least 2 points that are not at the average, not one but 2. If there was only one that was not at the average, it wouldn't make sense, would it? In fact, there has to be one that's to the right of the average a and one that's to the left of the average a. Let's call them, respectively, l and r. These are weights at those positions. Let's pick the one closest to the average a and move it to a, moving the other one in such a way so that the balancing point between them doesn't change. Here is l and here is r, and just these 2 balance in their middle, but a might not be in the direct middle. Let's say that a is a little bit closer to this one. I'll move this one to a, and I'll move this one the same distance, like that. This one is now at a because it was closer to a than this one was, but they both moved exactly the same distance toward each other. What's happened? We've only changed 2 out of our maybe a trillion values, and the sum of these 2 has stayed constant because we've done it in a symmetrical way. Capital S, our sum, has not changed. But one of these values became equal to a, and the other one got closer to the other value. If you just look at these 2 numbers, they had a fixed sum, just the 2 of them. Their sum is the same, but they're closer to each other in value, and so by the squarer is better property, their product has increased.

These are the only 2 values that we've been messing around with, and their product has increased. Therefore, the product of all n, which could be a trillion, the product of all of these numbers has gone up. By the squarer is better principle, we've been able to increase our product. The product of all n values has increased. We can keep on doing this. Our algorithm is that if they're not all at the average, 2 of them aren't. Then start sliding one to the average, the other one towards that one, [and] we'll keep increasing the product. Each time, we're using that squarer is better principle, which is just a 2-dimensional principle. It won't stop until all the values are equal to that average a because if one of them gets to the average and there's only one left over, that's an absurdity. That could never happen. At the very last move, we're going to do something like this, boom, right at the average, and it will join all of its brothers and sisters at the average. You can think of this as truly a simple, mechanical proof of the arithmetic-geometric mean inequality, which is quite a difficult thing to do algebraically.

Again, just to remind yourself, try to prove it in 3 dimensions algebraically or even in 7 dimensions. It will give you respect for this algorithmic method, and by the way, this algorithmic method is relatively unknown in American schools. This inequality is usually proven, if it's proven at all, using other methods that do not really explain why it's working the way that it does. Once again, this is something that comes, as far as I know it, from an Eastern European tradition.

Now we're ready to go back to this IMO problem. Remember the problem: We have a set of positive integers that sum to 1976, and we want to maximize their product. We've seen how powerful symmetry is, the squarer is better principle. Our instinct is that equal values optimize products, and this instinct has now been rigorously proven. But what if we're unable to make the parts equal? This, unfortunately, is the problem we have with this 1976 problem. There are 2 difficulties. We are not told how many numbers are in the product. We know that 2 wasn't quite good enough. We know that 3 improved it, and we know that 4 improved it still. We don't know how many numbers should be in the product, and even if we knew, we cannot guarantee that they would all be integers if we made them all equal. For example, suppose that the product were to have 10 numbers in it. Then, the average value is 197.6. It's not an integer, so we can't, in principle, make them all equal.

We know how to investigate this problem, though. It's a scary problem, but we know what to do. Let's get our hands dirty and make it easier. Is there anything special about 1976? Yes, that was the year the International Math Olympiad took place, where this problem was presented. It was in Austria that year. We certainly can replace 1976 with something smaller. Why don't we try 10? That's a reasonable number. If you look at 10, we could break 10 into $2 \times 2 \times 3 \times 3$, and that gives us 36. That seems to be the biggest possibility. When S is 20, we could get $2 \times 3 \times 3 \times 3 \times 3 \times 3 \times 3$, and that seems to be the biggest if we looked at all of the possibilities. Getting back to 10, remember, we could have just broken it into 5×5, which is 25. Breaking it into 3 parts, we could go $3 \times 3 \times 1$, which is only 9. Breaking it into 4 parts, $2 \times 2 \times 3 \times 3$, seemed to really do the trick. We can do a lot of experimentation, and if you made a table of these values, you fairly quickly make a conjecture that the optimal breakdowns only use the number 3 and the number 2 in them. The question, though, is: Why? Once again, we can solve this using an algorithmic approach, just like we did with the arithmetic-geometric mean inequality.

Assume to the contrary that we've broken down a number maximally. In other words, we've broken it down so that the product is maximal. The number is an integer, and we've broken it into integers, positive integers, that add up to this fixed sum. We've done it in as maximal a way as possible; the product is as large as can be. Assume to the contrary that we have numbers in it that aren't necessarily all 3s and 2s. Clearly, you can't have a 1 in your product, right? Why? Suppose you had, say, a 1 and then a 63. You could just take that 1 and the 63 and put them together, and you'd have the number 64, which is bigger. The number 1 isn't going to help because 1×63, the product is 63, but just merging the 1 with the 63 to get 64, instantly the product is bigger. You'll never have 1s. You already knew that 1s just don't help you here. But why only 3s and 2s? Consider: Suppose you had a 5 in your product. What could you do with the 5? You could break it into a 2 and a 3, and then its product is 6. It's better. What if you had a 4? Then you could turn it into a 2 and a 2, and its product hasn't gotten any worse. So 4s aren't bad, but you can always replace them with 2 2s. What if you have a 6? You could turn a 6 into a 3×3, which is 9, and 9 is better than 6. Is it general? Of course, it's general. Suppose you have a number that's not a 3. If it's a 4, we know what to do with it, and if it's anything else, just break it down. For example, if you have some number n bigger than or equal to 7—we've looked at 5 and 6 already—just turn it into 3 and $n - 3$ and note that $3(n - 3)$ is always going to be bigger than n as long as n is greater than 4. That's a very, very simple bit of algebra.

We can always take any number—for example, if the number is 100, at the very least, we can chip off a 3 and replace 100 with 3×97. Then take that 97 and replace it into 3×94, and just keep chipping off 3s until there's very little left. When we get to the end, there might be some 2s left. For example, if you start with 10, we'll turn 10 into 3×7. Then, we'll turn 7 into 3×4, and then we'll turn 4 into 2×2. We're left with $(3 \times 3)(2 \times 2)$. If you have more than 2 3s, though—you never need to have more than 2 3s because if you had 3 2s, you can replace them with 2 3s. In other words, $2 \times 2 \times 2$ is not as good as 3×3. Our algorithm is: Keep breaking it down into 3s, and then make sure that what's left over is as few 2s as possible. You might only have one 2, or you might have 2 2s, but you'll never have more than 2 2s because if you have 3 2s, you'd replace them with 2 3s. For example, 12 would break into $3 \times 3 \times 3 \times 3$, but 13 won't work that way. We'll take away a 3, and then another 3, and then another 3. That leaves us with 4 left, so 13 would break into $3 \times 3 \times 3 \times 2 \times 2$, and the final answer to our question is that 1976 would break into $3 \times 3 \times 3$ 658 times with 1 2 left over. The final answer is $3^{658} \times 2^2$, which is quite bigger than any of our initial experiments.

This almost equality problem has an almost equal solution, as your intuition would dictate, but why is the number 3 so significant? I was thrilled when a former student of mine, who's now a math professor, in the first week of my problem-solving course almost immediately suggested a completely crazy thing: to not worry about numbers not being integers. She said, "The values should be equal, so let's assume that they are." Call this common value x. Then, there will be $1976/x$ numbers in our product, since the sum of the x's is 1976. For example, if x were equal to 4, then there would be 494 numbers, all equal to 4. The product would be 4^{494}, which is already pretty big. What our problem is is to find the value of x which makes $x^{(1976/x)}$ maximum.

This is a terribly hard problem unless you know calculus, in which case, it's a fairly standard problem, but even if you don't know calculus, you can graph it. Let's replace the value of 1976 with a slightly smaller number, 30, so my computer doesn't have too much trouble. When you look at that, you find that the optimum value of x is between 2 and 3. If we try a different value, say, 189, we again get an optimum value between 2 and 3. The optimum product that we obtain is incredibly large. When you do standard calculus methods to this, no matter what the value of S is, whether it's 1976, or 30, or anything, you'll always get the value of e, 2.718 approximately. [This value] e is a very important number in calculus; it has about the same role in calculus as pi does in geometry, and it's quite a common number to come up in an optimization problem. What my student said, after a very quick calculation, [was] "The parts should all have size e." But of course, that's crazy because if the parts have size e, then the number of parts isn't an integer. What she was proposing was that we should have 726.93 numbers, each equal to 2.718. That actually gives a bigger product than the answer we got earlier, and of course, it's completely crazy. It's crazy because the problem asks for integers, and she explicitly ignored that.

Of course, it's not really crazy. It's really a great example of chainsawing the giraffe. The problem was an integer problem, and she said, "I want to do calculus, so I'm going to let the problem be continuous. I'm going to imagine that instead of integers, what if you now said, how about you're allowed to have the numbers have one decimal point, like numbers like 6.5?" How about 2 decimal points? What if we just added more and more decimal points? What we would converge to is her optimal solution, where the part size is e. It forced us, even if we knew we were using calculus illegally, to focus on looking at numbers near e, namely, 2 and 3, and that, of course, led to a rigorous solution.

Different people have different mathematical tastes. Some of you may feel that in this lecture we spent too much time proving these obvious facts—that squarer is better—but there are important lessons here. There's a big different between an intuitive hunch and a rigorous argument. We saw how we could prove an arbitrarily complex statement in n variables by rigorously understanding the 2-variable case, the squarer is better principle, and then ramping it up to higher and higher numbers of variables. This idea of large structures containing smaller ones is something we've seen before, for example, in Ramsey theory, and we most definitely will see it again.

Lecture Seventeen

Using Physical Intuition—and Imagination

Scope:

This lecture is inspired by a problem that I proposed for the USA Mathematical Olympiad only to find out that a variant of it had been used for years as a job interview question for a hedge fund. This challenging problem about marbles on a track combines much of what we have studied: invariants, symmetry, drawing pictures, and getting your hands dirty. Notably, we use physical intuition to get to the heart of the problem.

Outline

I. The marbles on a track problem: Several marbles are placed on a circular track of circumference 1 meter. The width of the track and the radii of the marbles are negligible. Each marble is randomly given an orientation, clockwise or counterclockwise. At time zero, each marble begins to travel with speed 1 meter per minute, where the direction of travel depends on the orientation. Whenever 2 marbles collide, they bounce back with no change in speed, obeying the laws of inelastic collision. What can you say about the possible locations of the marbles after 1 minute with respect to their original positions? There are 3 factors to consider: the number of marbles, their initial locations, and their initial orientations.

- **A.** This is a challenging problem, and we need a good venue for investigation.
- **B.** The geometry of the circle is irrelevant; only the fact that there is no beginning or end is important.

II. Hence the following warm-up, Martin Gardner's classic airplane problem. Several planes are based on a small island. The tank of each plane holds just enough fuel to take it halfway around the world. Fuel can be transferred from the tank of 1 plane to the tank of another while the planes are in flight. The only source of fuel is on the island, and we assume that there is no time lost in refueling either in the air or on the ground. What is the smallest number of planes that will ensure the flight of 1 plane around the world on a great circle, assuming that the planes have the same constant speed and rate of fuel consumption and that all planes return safely to the base?

A. Key idea: Circular motion can be modeled on a distance-time graph; just remember that the start and end are really the same point.

B. We will show that it is possible to fly around the world with just 3 planes. Call the planes *A*, *B*, and *C*.

1. First *A*, *B*, and *C* leave together, flying for 1 unit (1/8 of the way around the world).
2. Then *C* transfers 1 unit each to the other 2 planes. This gives *C* enough to return to base, while the other 2 now have full tanks.
3. *A* and *B* travel for 1 more unit, and then *B* transfers 1 unit to *A*. This leaves *A* with a full tank and *B* with enough to return to base.
4. *A* now has enough fuel to get to within 2 units of the destination. As soon as *B* returns to base (when *A* has reached the halfway point, with 2 units of fuel left), *B* refuels and heads for *A*'s location (traveling backward this time).
5. When *B* reaches *A*, *B* transfers 1 unit to *A*, and they both fly toward base. Meanwhile, *C* heads out again, reaching *A* and *B*, transferring 1 unit to each of them, and then all 3 head home.

C. Notice how the second half of the story is symmetrical with respect to the first.

III. Here is another warm-up problem, one that uses reflection: A laser strikes mirror *BC* at point *C*. The beam continues its path, bouncing off mirrors *AB* and *BC* according to the rule angle of incidence equals angle of reflection. If $AB = BC$, determine the number of times the beam will bounce off the 2 line segments (including the first bounce, at *C*).

A. Key insight: The broken-line path of the laser will be unbroken if you reflect across the mirror.

B. If it works once, it can be done again. Relentless reflection solves the problem by straightening out the path.

C. Instead of the actual laser path, look at the straight line. We merely need to count the number of times it intersects the boundary of one of the reflected mirrors. Answer: 6 bounces.

IV. Back to the marble problem. It is possible to find starting positions that change after 1 minute.

A. In one example, we start with 5 balls, and they end up in different positions.

1. However, the ending locations are a permutation of the original locations. For example, the black ball is now at the starting position of the blue ball, and the blue ball is where the green ball was.
2. The principle of reflection and the clever use of a simple distance-time representation on graph paper help us see that the problem is not that complex.
3. Suppose we were color-blind. A diagram of the paths would be the same as before, but we could not keep track of which marble is which. Pretend that the marbles are ghosts that can pass through one another. Then, of course, each marble ends up at exactly the same spot where it began.

B. This explains why the final positions of the marbles must coincide with the original positions, up to a permutation. Which permutations are possible? Remember that marbles cannot actually pass through one another, so the order of the marbles cannot change. All that can happen is a cyclic permutation.

C. How do we predict the permutation? If we started with 6 balls, there are 6 possible permutations. However, there are $2 \times 2 \times 2 \times 2 \times 2 \times 2$, which equals 64, different choices of initial orientation for the balls. How do these orientations influence the final result?

D. An invariant saves the day: Collisions only happen between balls of opposite velocity, and when they collide, the 2 balls swap velocities (and paths). Thus the sum of the velocities is constant. This is the same as conservation of angular momentum.

E. In general, if the net clockwise excess is c, then the net clockwise travel will be c full rotations of the circle. The only cyclic permutation that accomplishes this is the one in which each ball moves c balls clockwise.

F. The key ideas we used were physical intuition, looking for invariants, and symmetry.

Suggested Reading:

Gardner, *Martin Gardner's Sixth Book*, chap. 4.

Kendig, *Sink or Float?*

Tanton, *Solve This.*

Problems:

1. A monk climbs a mountain. He starts at 8 am and reaches the summit at noon. He spends the night on the summit. The next morning, he leaves the summit at 8 am and descends by the same route that he used the day before, reaching the bottom at noon. Prove that there is a time between 8 am and noon at which the monk was at exactly the same spot on the mountain on both days. (Notice that we do not specify anything about the speed that the monk travels. For example, he could race at 1000 miles per hour for the first few minutes, then sit still for hours, then travel backward, etc. Nor does the monk have to travel at the same speeds when going up as going down.)

2. Imagine a laser beam that starts at the southwest corner of a square and moves northeast with a slope of 7/11. How many times will it bounce before it returns to its starting point?

Lecture Seventeen—Transcript
Using Physical Intuition—and Imagination

This lecture, like the earlier one about games, does not attempt to teach you new problem-solving strategies or tactics. Instead, we will look at a few problems in depth and use methods that we've already seen. Our goal is to continue to reinforce the all-important ideas of symmetry, but now we'll add its "mother," invariance, now that we've learned something about that. In particular, we will channel our intuition about the so-called "real world" to use physical invariants to solve a problem. We'll also discuss the process of problem solving in more detail than we have recently because I personally found the focal problem of this lecture to be rather difficult. It took me over a week to solve it. I want to remind you that what you're seeing here are, to some degree, polished lecture solutions, where you're not seeing the process of making the sausage, as it were. You have to remember, as you know by now from all the problems that you've been working on, that often it takes a long time to get started. The investigation can go in many, many different directions. That process of problem solving is really a complicated one. It's important to remember that—that what I present in these lectures is not necessarily the way I thought of it originally. Often the way I thought of it originally took a lot more time, had a lot more stupidity, had a lot more cursing, and just lots and lots of meandering. Of course, seeing all that meandering can be pretty boring, so I'm trying to spare you some of those details, but I'll mention a few in this lecture.

Our focal problem is a physical one. It involves colliding marbles on a circular track. Imagine a circular track; say you're looking down at a track. It's like a skating rink, but it's just the circular perimeter, and the marbles are constrained just to go around the perimeter of the track. The width of the track and the radii of the marbles are negligible. The marbles are—think of them as teeny, tiny, solid balls. Each marble is randomly given a direction, an orientation, whether they're going clockwise, if you're looking at the circle from above, or perhaps counterclockwise, and the marbles are all traveling at the same speed. The circumference of the circle is a meter, and the marbles are traveling at a speed [of] 1 meter per minute. Again, imagine these marbles are on this circle, and some of them are traveling one way and some are traveling the other way. Well, you know what's going to happen; marbles are going to collide. When they collide, they bounce back with no change in speed. They obey the laws of inelastic collision, but remember, they're constrained to this circular track. Here's the question: What can you say about the possible location of the marbles after 1 minute with respect to their original positions? The marbles start out at some random positions; they all are going one circumference per minute. The circumference is 1 meter, and they're going 1 meter per minute. Some of them are going clockwise, and some of them are going counterclockwise, so some of them might bounce into each other. The question is: What would happen after a minute? There are 3 factors to consider—the number of marbles—we could have 2, 3, a billion; the initial positions of the marbles; and also their initial directions, whether they're counterclockwise or clockwise. There are lots and lots of possibilities.

This is a pretty hard problem, a pretty complicated problem, and I originally came up with this idea. The circular marbles on a track, certainly I'm not the first person to ever imagine this, but I created this problem involving these distances and I submitted it as a problem for the USA Math Olympiad, which I sometimes submit problems for. We decided to reject the problem because one of the people on the committee worked for a hedge fund, and he used an interview question that he thought was fairly similar to this problem. He felt that if it was a job interview question, the basic idea might be too well known, and so it wasn't a good idea to use in this competition. That doesn't matter; what matters is it's an interesting problem and it's a challenging problem.

The question is: How do we investigate it? All problems need an entry point for investigation, and one of the difficulties here is that circles are hard to draw. I don't know if you're like me, but I cannot draw a circle very well. We could make a checklist of various things to try. For example, the extreme principle, but there are no values that we can talk about bigger or smaller. Parity doesn't seem like a good idea. Wishful thinking—we don't even know what to wish for yet. The problem is just getting started.

The first breakthrough is to realize that the geometry of a circle is irrelevant. The fact about a circle that's important is the old saying that a circle has no beginning or no end. How can we imagine a circle in a way that's easier to draw? This is where experience is sometimes important. Sometimes someone has to tell you, "Here's a way to think about a circle," or sometimes easier problems give you more of an inspiration. A real inspiration for this is playing video games, but rather than talk about video games, let's first look at a somewhat easier problem. This is a great classic from Martin Gardner.

Several planes are based on a small island; the tank of each plane holds just enough fuel to take it halfway around the world. The fuel can be transferred from the tank of one plane to the tank of another while the planes are in flight, and it takes no time at all, so you can instantly transfer fuel. The only source of fuel is on the island; we assume there's no time lost in refueling in the air or even on the ground. What's the smallest number of planes that you need to ensure the flight of one plane around the Earth, around the world in a great circle path, assuming that the planes all have the same constant speed, the rate of fuel consumption is the same, and of course, that all the planes return safely back to their base? Well, I've given this problem to my students many times, and they have struggled because they've tried to draw circles. They realize they don't need to draw a sphere. The problem involves some kind of circular motion, but just drawing this on circles drives them crazy. Every once in a while, maybe one in 5 students will think about the fact that you don't need a circle. All you need is a line whose beginning and endpoint are the same. For example, if you have a distance-time graph, if distance is vertical and time is horizontal, we could let the top and bottom be the starting point and ending point of our circle. In other words, they're the same point. If you look at this graph here, this shows a plane traveling completely around the world. The problem, of course, is the plane doesn't have enough fuel. It only has enough fuel to get halfway across. I've divided time into 8 units and also the circular length into 8 units, and again, we're using this come-back-to-your-beginning idea: If you can think of the circle as you start at 0 and you go up to 8, you're dividing the world into 8 sections, so you go 0, 1, 2, 3, 4, 5, 6, 7, and then 8 is back to 0.

Imagine now that we have a fleet of 3 planes. Clearly, one plane won't do the trick; let's try 3 planes. Let's color them black, green, and red, and so they all leave together. What you're seeing there is a graph of all the planes traveling together if they could. Of course, they're going to run out of fuel, but let's just see what happens after 1 unit of time. After 1 unit of time, the planes each have 3 units of gas left, 3 units of fuel, because there are 8 units of distance, but you can only go halfway around the world, so you have only 4 units of fuel. After 1 unit of time, everybody's lost a unit fuel, which is bad, but now what we'll do is we'll have a fuel transfer. We'll let the black plane be the plane that will try to get around the world, and so what we'll do is have the red plane give up its fuel, give up all but 1 unit of fuel, to its 2 comrades. At this point, instantly, now the black and the green both have 4 units of fuel, but the poor little red plane only has 1. Now what it does is it heads back to base because otherwise it's going to crash, and it has just 1 unit of time to do it. It makes it back to base in time. Now after 2 units of time, we have the 2 planes, the black and the green, and now they each have 3 units of fuel. Well, they're going to run out of fuel unless something else can happen, but now what happens is the green plane decides to sacrifice its fuel and give it to its friend the black plane. The green plane gives away 1 unit, bringing the black plane back up to 4, and then the green plane has 2 units. That's just enough for it to get home and survive, and now the black plane has 4 units, and it travels for the next 2 time periods. Now the great thing is it has made it halfway across the world, and it's not crashing into the ocean. It has 2 units of fuel left, but of course, that's not enough to get back to the other end of the world, to get around the full circle, because the plane needs 2 more units of fuel. The situation is a symmetrical one because what can happen now is the red and green planes can now start flying toward the black plane and meet the black plane in 2 more time units, transferring the fuel and repeating the process just going backwards, and so we have this beautiful symmetrical situation that requires exactly 3 planes.

It's a beautiful solution, and it would be almost impossible to figure it out if you didn't have something like graph paper with this notion of when you get to the top of the graph paper you're back to the beginning of your trip, the idea of converting a circle into a line with the understanding that the 2 ends of the line are linked together. Now, there's a question: Is this the optimal solution? Could you have done it with 2 planes? Well, I'm not going to answer that question right now. I'd rather that you think about it, but it is an interesting exercise to see if you can get around the world with 2 planes. At the very least, we know we can do it with 3 planes. When I've given this problem to my students, the average number of planes that they come up with is usually around 4 or 5, sometimes even 8. Those are people that really have trouble drawing their circles. Three is really an unexpected answer.

Let's go back to the marbles problem. The idea of Martin Gardner's airplane problem teaches us [that] a good venue for investigating any kind of circular motion is to throw circles away and just think of a circle as a straight line whose 2 ends are glued together, in some sense. We'll use graph paper; we'll use a distance-time graph, just like we did with the airplanes, only it's a little more complicated. But we can get our hands dirty finally; that's the important thing. Here's a simple example. Again, we'll let the vertical axis be the clockwise position, and we'll go from 0 to 1. So 0 and 1 are the same point, even though there's a 0 at the bottom and a 1 on the top. You can think of the graph paper as maybe wrapped around itself in some strange way, sort of like a video game. If you're playing a video game, you go off to one end, and then suddenly you come back to the beginning. It's the same exact idea. In fact, if you do it in the vertical direction, you're also wrapping around, which means you're playing on a torus, but that's a whole other story. Let's get back to our problem, which is complicated enough.

The horizontal axis is timed in minutes, going from 0 to 1 in both cases. Thus, a single marble could be depicted with a line with a slope of 1 because it's going at unit speed. In this picture, we started with a red ball that's going clockwise and a black one that starts half a circle away going counterclockwise, and so the 2 marbles collide. Of course, when they collide, they go in opposite directions, so the color is swapped. Notice the black path turns around, and the red path turns around. The collision will happen in a quarter of a minute, and then after another 15 seconds, another quarter of a minute, the red ball is back where it started at the beginning of the circle, at some point labeled 0 on the circle. Now it's continuing in a counterclockwise direction, and so if you look at where it continues on its path, it looks as if it's jumping to the top of the path, but it's really just continuing because remember the top of the path is exactly the same as the bottom of the path. It's continuing in that counterclockwise direction, so it's going down. Then it eventually collides with the black marble, and once again, we have a color swap.

It's hard thinking about what the marbles are doing, but the beautiful thing about the graph paper is that you don't have to think. All you're doing is drawing colored lines, and when 2 colors meet, there's a collision and the colors change. In this case, the 2 marbles ended up exactly where they were. The black one was in the middle, and the red one was at position 0. It's just a simple example, but it gives us a good conjecture that perhaps nothing at all changes. Perhaps after 1 minute, everything is where it belonged, everything is where it started off. Perhaps it's just a simple invariant of that kind.

Let's hold off on this for a moment because we've only done one experiment, and let's get a little more practice involving the idea of marbles colliding with one another because what we're also doing is having lines reflecting off each other.

Let's look at another problem that involves reflection. This is another problem that I submitted for a math contest. This one actually was accepted, and it was for the feeder exam before the USA Math Olympiad. I'm adapting the problem a little bit, but the basic idea is that you have a laser beam being shot at 2 mirrors that form an angle. The laser beam is bouncing off the edges, and the beam comes in and goes bounce, bounce, bounce, bounce, bounce. Eventually, it will bounce out. The question is: Based on the angle of incidence of the original beam and the angle that the mirror makes, what will the number of bounces be? Let's investigate this with a picture. Here's the red laser beam coming in, and clearly, it's going to make at least a few bounces. Our rule for bounce is the normal physical rule that angle of incidence equals angle of reflection. If you don't like laser beams, you could think of this as billiards, although again, they are infinitesimal billiards or marbles, just like our marble problem. We have reflection going on. What we want to do is find a way to use the picture to count the number of bounces.

You've seen problems like this before. You know from your knowledge of symmetry that imposing as much symmetry as possible is important, and here, the whole theme of reflection is screaming at you to try some more reflection. Just like the problem with fetching water for your grandmother, the natural thing to do is to reflect the picture itself, and when we do that, we have our red beam reflected into a green beam now. You can see that we can take the original path, and where it does its first bounce, instead, follow it along onto the lower path. Thus, what was a broken line becomes a straight line, at least the first 2 segments of the path. We can keep on doing this; nothing says we can't reflect more than once. Why don't we just reflect as much as we want? If we keep reflecting in order to straighten out the path, eventually, what we'll do is we'll take that original angle of the mirrors and reflect as much as we can to form as much of a circle. Then the problem is just a matter of counting how many intersections with the sort of wedge boundaries that red line has to take before it exits the circle. If you look at the picture, we have 1, 2, 3, 4, 5, 6 intersections; that corresponds to 6 bounces. Sure enough, if we draw a very, very careful diagram, we can see that we do have 6 bounces before the laser leaves the mirrored angle forever. The moral of that story—this idea of reflection as much as you need and focus on turning a broken path into a straight-line path is an inspiration for us. We can imply it back with the marble problem. Remember, we conjectured that there's no change—that the marbles end up exactly where they were.

Here's a much more complicated example using 6 different balls. I just use a number of different colors. If you carefully follow the paths, the rule, again, is quite simple. Whenever 2 marbles intersect, the colors just swap. If we continue the path, remember when you get to the top of the graph paper, you then have to go back to the bottom. If you look at this picture, again, if you look carefully, you'll see that the marbles are exactly in the same places as where they started; nothing has changed. Could the problem be that boring? No, it's not quite that boring. Here's a slightly less complicated example where I removed the purple ball, and now if you look at the colors, there are 5 balls, but the starting position of the balls is different from their ending position. The balls moved. Now there's something scary about the problem. What happened? When I first was investigating this, I said, "Oh yeah, the balls all stay where they were." Then I realized, "No, they don't." I needed to figure out where these balls ended up. Here's a nice insight. You'll notice the location of the balls have changed. In other words, the green ball has ended up differently from where it started, but the green ball ended where another ball began. If you look at where the balls ended up, they might be different from where they were, but they're still places where the balls started out. In other words, what's happened, in some way, is that the ending position of the marbles is some sort of permutation, some sort of rearrangement, of their original positions. If you were colorblind and you looked at the beginning and the end, they'd look the same. That's a good way to think about it. Why is that true?

Here's a simple way to do it. Let's use, in a sense, an element of wishful thinking. The difficulty of this problem is that the marbles bounce into each other, and then they collide and they change directions. That's complicated physics. What if it didn't happen? What if the marbles were ghosts? What if they just traveled through each other? While we're at it, what if we were colorblind? Then we would just have a picture that looked like this, where we just have our paths go, and of course, the marbles are going to end up exactly where they started because they travel one circumference in one unit. That's a simple proof that the end result has to be a permutation or rearrangement of the original colors.

What we need to figure out is which permutation happens. The problem is when the marbles are traveling in reality, they will bounce into one another. We have to figure out how the permutation occurs. The big question is: Which permutations are possible? How many ways could this happen? For example, you might have just 3 marbles, and they could all be traveling in the same orientation. Then there's no collision at all, and of course, the marbles end up where they started. But it's possible for the 3 marbles to be permuted in several different ways. You know from an earlier lecture that it's possible to permute 3 different things, to rearrange their order, in 6 different ways. Could there be 6 possible solutions? Well, I struggled with this for days, and days, and days. I'm a mathematician, and I tend to forget about the real world, which is a mistake because the real world informs the mathematical world. One morning I was in the shower, which is where I do a lot of thinking, and I waste a lot of time because I usually forget about the passage of time. One morning I was taking my 45-minute shower, and it occurred to me. I said, "Man, I'm stupid," because I realized that I had thought of this beautiful idea of marbles passing through each other, this idea of ghost marbles, that allowed me to realize it was a permutation, but then I forgot that they're real marbles. Marbles don't actually go through each other. For example, here's a little red marble, and it's traveling behind a little blue marble. The red marble will always have a blue marble next to it, and likewise, if there's a green marble behind a red marble, the order will always be green, red, blue because the marbles can't pass through each other. If there's going to be a permutation, it has to be a cyclic permutation, in other words, just a permutation where we don't scramble things around but we just sort of move it as if it's a necklace, like the necklace analysis we used for Fermat's little theorem.

If there are 6 marbles, there are only 6 possible cyclic permutations of them. How do we predict which permutation it will be? We've reduced the problem to a simpler situation. When I was working on this personally, I remember spending hours watching *Star Trek*. I know this sounds nerdy, but I actually was watching *Star Trek* with my graph paper in front of me and my colored pencils. I would just do experiment after experiment and look at the different possible cyclic permutations. The way I measured them was by how much the marbles turned. Like if we started with red, blue, green, and the red moved to the blue, the blue moved to the green, and the green moved back to the red, that would just be a single turn. You could think of 2 turns. And with 3 marbles, 3 turns would just get you back to the exact original position. I did experiment, after experiment, after experiment, and I'm not making this up, until another shower came. I shower more often than this, but another mathematical shower came. I finally figured out what was happening. I was thinking to myself, well, I'm a problem solver. I know that invariants are important in almost any mathematical situation, and I was trying to think of mathematical invariants, but I wasn't thinking of physical invariants.

Let's think about physical invariants because it's really quite simple. What stays the same at all times in this problem, not at the beginning and at the end, but at all times? Not the velocities of each ball since when a ball collides with another its velocity changes sign. I say "sign," but I just mean direction, like you can think of it as positive and negative, which is what's happening in the graph. The graph going up means you're going clockwise, and going down means you're going counterclockwise. If a clockwise ball hits a counterclockwise ball, then they switch directions, and so the velocities do change sign. The velocities are always either ± 1. The 2 balls swap their velocities. They also swap their paths, but the important thing is that the sum of these velocities is constant because when a collision happens, the 2 numbers just switch position. They're the same 2 numbers, and if we were to add up all velocities, we're always going to have to same number, always. This is really conservation of angular momentum, and that's just a fancy term. A really simple way to think of about is just [that] the sum of the velocities at any given time is constant. It allows us now to think about the marbles as a collective. For example, here's the example: 3

of the balls are going clockwise, and 2 are going counterclockwise. If you want to put signs on this, the net motion would be $+2 + (-3) = -1$, in other words, 1 unit clockwise.

Collectively, the entire set of balls is moving clockwise 1 meter per minute. The only way that can happen is if each ball then moves to its clockwise neighbor. If the net clockwise excess is some number *c*—*c* for clockwise—then the net clockwise travel will be *c* full rotations of the circle. The only cyclic permutation that accomplishes this is the one in which each ball moves *c* balls clockwise to its neighbor. Let's look at a simple example to make this clear. This is what has to happen if the value of c is $+1$. Here's a situation with just 3 marbles. What that means is 2 of them are going clockwise and one of them is going counterclockwise. Collectively, the entire group is going to be traveling one full circumference. How do you make that happen? If *c* moves to *a*, and if *a* moves to *b*, and if *b* moves to *c*, then collectively, that is one full rotation. That's the example of $c = 1$. If $c = 2$, then the marble labeled *c* would go all the way to *b*, *b* would go all the way to *a*, and a would go all the way to *c*. Collectively, that would be a turn of 2 full rotations, 2 full circumferences.

As I said before, this problem took me a long time. It was definitely, using my notion of time scale, a level-4 problem, but I didn't give up and although I felt stupid a lot of the time, that didn't mean that I had didn't have a good time. Remember, most mathematicians spend most of their time feeling kind of stupid but usually being happy about it. The real moral of the story was the extreme importance of using physical intuition, the idea of sort of conservation of sums of velocities, and the elementary fact that eluded me for a long time that 2 physical balls cannot actually pass through one another. These things are very, very helpful, and they make a mathematical investigation a lot of fun. It's even better if we can combine this reality-grounded intuition with imagination that transcends reality, and we use that with these ideas of ghost paths. Remember, the physical world is important, but the mathematical world is a world where imagination plays the most important role. In the next lecture, we will leave the physical world behind, and we'll focus on plane geometry, but we'll take a very fresh and imaginative perspective, one where the notion of invariance, now our old friend, plays a very central role.

Lecture Eighteen

Geometry and the Transformation Tactic

Scope:

This is the only lecture in the course wholly devoted to geometry. We look at geometric transformation, a great example of the problem-solving strategy of reversing one's point of view. We apply this idea to a number of problems that initially appear completely intractable but become almost trivial once we are comfortable with dynamic entities like rotations, vectors, and reflections.

Outline

I. We devote this lecture to a tiny fraction of transformational geometry because of its connections to ideas already familiar to us, in particular, symmetry and invariants.

- **A.** Before we get into the details of transformational geometry, here is a problem to think about that we will solve later with the transformational methods we develop.
- **B.** Suppose a pentagon (not necessarily regular) is drawn on the plane. The midpoints of each side are found. Then, suppose the original pentagon is erased, leaving only the midpoints. Can the original pentagon be reconstructed?
- **C.** Transformational geometry was pioneered in 1872 by the German mathematician Felix Klein and is an example of the strategy of reversing one's point of view. Klein suggested that the proper way to think about geometry was not to focus on the objects but instead to contemplate the transformations that act on them.
- **D.** Why are transformations important? Because they are not just geometric but also algebraic entities.
- **E.** You do not add or subtract transformations, but you can sort of multiply them by composition.

II. In our exploration of the algebra of transformations, we restrict our attention to the plane and look at just 3 types of transformations: reflections, rotations, and translations.

III. Composition of reflections: Let F_h denote a reflection across line h. Note that F_h leaves all the points of h invariant. In general, the fixed points of a reflection are a line.

IV. When composing 2 reflections, there are 3 cases: The lines are the same, parallel, or meet in a single point.

V. Let's look at the composition of 2 rotations. First we need a lemma about what happens to a line under rotation.

A. Rotated line lemma: Suppose line h is rotated by a rotation with center A and angle α. Let h' be the image of h. Then h' makes an angle of α with h. The proof is a similar triangles argument.

B. Now that you have a feel for rotating lines, here is a fantastic example. Suppose you are given 3 parallel lines: ℓ_1, ℓ_2, and ℓ_3. Is it possible to construct an equilateral triangle such that each vertex of the triangle lies on one of each of these lines?

1. Starting with A, what transformation leaves parts of the triangle invariant?
2. One idea: Clockwise rotation by 60° about A.
3. Now consider this rotation, but let line ℓ_3 go along for the ride!
4. Its image is ℓ_4, and where it intersects ℓ_2 is the point B. Now we can construct the triangle.

C. By using parts of the figure and the knowledge that the rotation moved one point of the unknown triangle to another unknown point, we were able to construct the location of both points!

D. Now we are ready to understand the composition of 2 rotations.

1. No matter where the centers are, the composition is just another rotation about this center, where the angle is just the sum of the 2 angles.
2. But if the angles for 2 rotations add up to 360°, then the composition of the 2 rotations is a translation.

3. This is an unexpected but helpful fact that is the tool we need for the pentagon problem. Let's call it the reflection-translation tool.

VI. Now let's use this tool to solve the pentagon problem that we began our lecture with.

A. Use wishful thinking to pretend you know the pentagon's vertices: A, B, C, D, and E. In reality, the only points that we know about are the midpoints: X, Y, Z, V, and W.

B. Crux idea: Rotations of 180° about midpoints. These rotations bring (unknown) vertices to vertices.

C. Compose the 5 rotations by 180° about X, Y, Z, V, and W; this brings A back to its starting point.

D. So do one more rotation of 180°, about X again. That brings A to B.

1. It was a composition of 6 rotations, each of 180°, and $6 \times 180° = 1080° = 3 \times 360°$.
2. So it is a translation!

E. We can perform this 6-rotation composition on any point we like. It took A to B. However, there is one problem: We do not know where A is!

F. Just pick a random point P. Perform the 6-rotation composition on P, getting successive points $P_1, P_2, \ldots, P_6$.

G. We know that the vector from P to P_6 is the same as the vector from A to B!

H. The line segment joining P_1 and P_6 has the same length as the mystery side AB of our pentagon. Draw a line parallel to this segment that goes through X, and mark off equal segments on either side of X whose length is half of AB. We have just reconstructed segment AB!

VII. Why did it work? The higher-level reason for this is that we did not use the transformations randomly, but instead searched for transformations that leave parts of our problem invariant.

Suggested Reading:

Liu, *Hungarian Problem Book III.*

Needham, *Visual Complex Analysis*, chap. 1.

Yaglom, *Geometric Transformations I.*

Zeitz, *The Art and Craft of Problem Solving*, sec. 8.5.

Problems:

1. Let *IJK* be an arbitrary triangle with equilateral triangles constructed on each edge. Thus *IJL*, *KJM*, and *IKN* are all equilateral triangles. Prove that *IM*, *KL*, and *JN* have exactly the same length. (Hint: Perform a rotation or 2.)

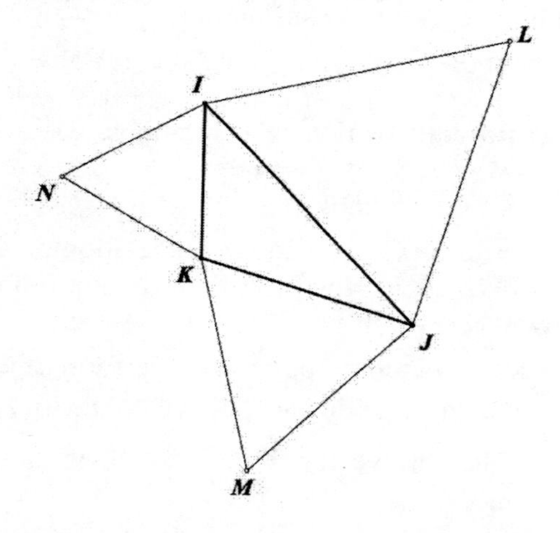

2. Let R_k denote rotation by 90° counterclockwise about the point $(k, 0)$ in the plane. The composition of R_0, R_1, R_2, R_3, in that order, has a total of 360° of rotation and hence is a translation (possibly the identity). What translation is it? Can you generalize to n rotations, where each is about $(k, 0)$, and the angle is $360/n$?

Lecture Eighteen—Transcript

Geometry and the Transformation Tactic

This lecture is a little different from the others in several ways. First of all, it's the only lecture that's exclusively concerned with geometry. In this lecture, when we see symmetry, it is most definitely not metaphorical, but instead, it's literal, visual symmetry. Second, we're going to look at a class of interesting problems that require a very special set of tools, so we will need to develop these tools briefly. For part of the lecture, we will explore several lemmas that may appear, at least temporarily, to be a little technical and strange, but don't worry. It will all make sense, and as we expect with exotic tools and tactics, the results will be dramatic and nearly magical. For the tools that we'll use, they're sort of a mathematical version of modern cancer drugs that are targeted for very specific cells. What we'll be developing with our transformation tactic is a way to craft transformations that target just parts of a geometric problem, leaving parts of the problem invariant and moving other parts of it around, as you'll see.

As I've said, this is the first and only lecture that's 100% geometry. Why is geometry so underrepresented? American school curriculum has very little geometry, and the geometry tends to deemphasize the old-fashioned Euclidean proof style. In contrast, Eastern Europe, of course, has a big tradition of geometric excellence, and the way that happens is by lots and lots of homework. I have a Hungarian friend who when she went home for Christmas vacation when she was 14 years old had 1000 construction exercises to do, 1000 for Christmas vacation. She said it was fun.

It's not a coincidence that in the IMO training program, the head instructors have, over the past 20 [to] 30 years, tended to be Romanian and Chinese with some Americans, but usually the leaders come from Eastern Europe and now from Asia. I wrote a book about problem solving about 10 years ago, and in the first edition, I was too scared to put in a geometry chapter. I finally did in the second edition, but I did call this chapter "Geometry for Americans." I did that because if I said "Geometry for Dummies," it would be a little bit insulting. Of course, nobody is a dummy here; it's just a matter of inexperience. Let's get a little experienced.

There's lots of geometry to study. In this lecture, we're just going to look at a very, very particular kind. There is what's called synthetic geometry, sort of old-fashioned Euclidean style, which is great stuff, but we're not going to do that. There's analytic geometry, which I tend to look down upon, even though it's very, very important for lots of applications, but it tends to lead to difficult algebra. What we will look at is transformational geometry. I've chosen it because of its unexpectedness and its very much higher-order connections to problem-solving ideas that we've already seen, such as symmetry and invariance. It's relatively unknown except at the highest levels of problem-solving studies, such as students preparing for the International Olympiad, but it's quite common for, say, a Bulgarian student to learn as a 14 year old.

Before we get into the details of transformational geometry, here's the key problem that I want you to think about. Suppose a pentagon, not necessarily regular, is drawn on the plane. Mark the midpoints of each side. Then erase everything, leaving only the midpoints. Can you reconstruct the original pentagon? In other words, you're just faced with 5 dots. What do you do? What is there to work with? This is a mysterious problem that we will solve using our new methods. If you think about it, it's quite a mystery. How do we find the old pentagon just from the midpoints? There's nothing to hold onto.

The idea of transformational geometry was pioneered in the 1800s by the German mathematician Felix Klein. What it is is a great example of the problem-solving strategy of flipping a point of view. What Klein proposed, in an important address in 1872, was that the proper way to think about geometry was not to focus on the objects of geometry, that is, like circles, triangles, etcetera, but instead, contemplate the transformations of geometry, in other words, things like rotations, translations, reflections, the things that act on the objects, and particularly, to study the ones that act at least in a partially invariant way. A geometric analogy is: Suppose you're looking at a ladder where every other rung is blue. You could pick a transformation of the ladder to translate it by 2 rungs, and that will take the blue rungs back into blue rungs. It will actually move the ladder, but it will leave part of the ladder—part of its color scheme will stay invariant. That's an example of the type of transformation that we're going to study. We've used this idea back when we discovered the tactic of imposing symmetry in a problem.

Klein's suggestion seems pretty innocent, but it actually was a profound suggestion that really changed mathematics and not just geometry because the notion of an object and a transformation can be generalized. Remember, when I say transformations, I don't just mean things that move geometry. What makes them useful here is that transformations are also algebraic entities, like polynomials. You can't add or subtract transformations, but what you can do is you can do a sort of multiplication that is called composition. Composition is like glue. Given one transformation, say, a rotation, and then another transformation, maybe another rotation, we can say: What happens if we do one after another? For example, suppose we have the transformation that takes the entire plane and rotates it by 90 degrees counterclockwise. Imagine that I'm standing on that plane, and the axis of rotation goes right through my center, right through my head. Let's call that transformation R. If we perform R once, this is what happens. If we perform R twice, I turn around, 3 Rs, 4 Rs. We don't say 4 Rs gets me back to the original. What we say is that R^4, R composed with R, composed with R, composed with R, brings me to the identity, to the transformation that does nothing. We can do a sort of algebra with transformations.

We'll need to look at a few more examples of this sort of thing before we can do hard problems, but first, let's take a look at a fairly easy warm-up problem that illustrates the idea of a transformation that acts in a focused way, sort of like a cancer drug. This is a very popular American math team question of the short answer/multiple choice tradition. It's a good way to look at this transformation tactic. Imagine you have an 8×8 square, and a right triangle is superimposed on top of it, where the right angle goes right through the center of the square. The right triangle is oriented so that it cuts the top end of the square into segments of length 2 and 6. It's an 8×8 square, and let's say the lengths of the legs of this right triangle are 5 × the square root of 5 and 4 × the square root of 5. The question is: What's the overlap area? What is the area of the shaded piece? It's a quadrilateral, and it's a little tricky to break it up into parts that you can measure, but the beautiful insight is to notice that if you take a perpendicular from the center of the square, going to the midpoints of the top and the right end of the square, we cut off these little triangles. There's the little red triangle and the little green triangle, and if you think about it, those triangles are congruent. They're congruent because the large triangle was a right angle, and each of these small triangles are also right triangles. They have equal lengths; XG and XI are equal. The angle that's cut off has to be the same because you're subtracting the gray angle (*GXH*) from a right angle in both cases. Those triangles are the same triangle, so their areas are the same. Rotate. When you rotate, suddenly, the overlapping area is a simple 4×4 square, but the real point is that no matter where we rotate that large right angle, the overlapping area will be invariant because the red triangle will always equal the green triangle. That's the idea.

Let's look briefly at some more of the algebra of transformations. This is the point in our lecture where we're going to do a few lemmas, but they're interesting lemmas. We're just going to scratch the surface of this topic, in a literal way. We're only going to restrict ourselves to the plane. There are many, many other places where we can do this kind of geometry, but for now, we'll just stick to the Euclidean plane. We're only to look at 3 types of transformations—reflections, which you know about; rotations, which you've seen; and translations, which are even simpler. All that a translation is is motion by a fixed vector. There's a translation. Of course, don't think of me as being 3-dimensional; just think of something happening on the plane. These are rigid motions; they don't change an object. These are done to the entire plane. If we talk about a rotation around the origin, think of the entire plane as moving by some angle.

What we're going to do is investigate what happens when we compose some of these. What happens when we compose 2 reflections? What happens when we compose 2 rotations? Let's start with reflections. We'll use the letter *F* for 'flection, because after all, reflection and rotation start with *R*. We'll use *F* for reflection, and we'll use a subscript for the line that we're reflecting across. Lines are usually lowercase, so F_h would be the transformation of the entire plane that just reflects everything across *h*, like a mirror. Now, F_h leaves all the points on the line *h* alone. These are the invariant points. These are also called fixed points, and for any reflection, the fixed points are a line. What happens when we compose 2 reflections? There are 3 cases. Either the lines are the same, which is boring, because then you just do the reflection twice, so that's the identity transformation, or the lines are parallel, or the lines meet in a single point. Let's suppose the lines are parallel; *h* and *m* are parallel. Let's imagine that we start with a random point *A*. Then we do the reflection across the line *h*. Then we do the reflection again across the line *m*. Our starting point *A* moves to *A′* when we reflect across *h*, then A' moves to A''. We call A' the image of A under the first reflection, and A'' is the image of A' under the second reflection. If you look at the picture, A moves to A', which moves to A'', and if you think about it, it's clearly just going to be a translation. It's very simple to see that it just sort of jumps across.

What if h and m intersect in a point, if those 2 lines intersect in a point? Focus on the intersection of the 2 lines, the intersection point X. When we do the reflection across h, X doesn't go anywhere because it's on the line h, but it's also on the line m, so when we do the second reflection, X doesn't go anywhere. The image of X under the composition is X; X is an invariant point. If you think about it, that's the only point on the plane that's going to be invariant. What kind of transformation will have a single fixed point? The only possibility is a rotation, but which one? One way to discover it is we start with a random point A. Reflect across h. It goes to A', and then take that point A', and reflect it again over m. It goes to A''. So A goes to A', to A'', and it looks as though point A has sort of rotated clockwise, but how do we know what the exact rotation is? Just use your imagination. We know it's going to be some rotation, and that random point A is not in a good place. Let's move it somewhere where things are more constrained. If you imagine moving A towards line h, then when we do our first reflection, it doesn't go anywhere. When we do our second reflection over m, it's going to reflect straight across line m. What you're doing is taking a point that's on line h and reflecting it across line m. If you draw a diagram of that, you'll see that the actual rotation is twice the angle between the lines. These are not lemmas you need to memorize; the idea is just to get a feel for how this kind of algebra is investigated.

Where we're going and what's most important for us is to look at the composition of 2 rotations. Here we will get to a lemma that you will want to remember. Before we do that, we need a sort of preparatory lemma about what happens to a single line when it is rotated. Here's the rotated line lemma. Suppose you take a line h and you rotate it by a rotation with the center of A and an angle of α. Every rotation has a center and an angle, and the convention is usually for angles to be Greek letters, not always, and the convention is that the rotation is counterclockwise. Imagine that h is rotated by some rotation with the center of A, which might not be on the line—it's just somewhere—and let h' be the image of h under this rotation. The lemma says that the angle that h' makes with its original line h is the same angle α.

Here's a picture of the situation where the value of α is 30 degrees. We take line h, and we rotate it about the point A, 30 degrees counterclockwise. You can see that it becomes the line h', and it also appears that the angle between those 2 lines is also 30 degrees. This is an easy thing to prove with really simple geometric arguments. The way to actually imagine rotating that line h would be to draw a perpendicular from A and think of that as kind of a swing arm. Here's the line, and we're kind of moving it like that. Then, draw a perpendicular to line h', as well. What you will notice in this diagram is that there are 2 right triangles, the large one with the red angle and the small one with the green angle, and they both share the point Y. They're both right-angled triangles, and so therefore, if they have angle Y and they have a right angle, what's left has to be the same. Therefore, that little red angle and the tiny green angle are the same, and the red angle was the rotation amount. That was our α, and that's a very simple visual proof that the lines will intersect and the angle between them will be the angle of the rotation.

Let's apply the idea of rotating lines to a fantastic example. Suppose you're given 3 parallel lines, $L1$, $L2$, and $L3$. Is it possible to construct an equilateral triangle such that each vertex of the triangle lies on each of those 3 lines? Here's an example using wishful thinking. We have our lines $L1$, $L2$, and $L3$, and there's a beautiful equilateral triangle. The problem is we don't know how to put it there. Let's just put A somewhere on the bottom line, but how do we find B and C? You could do it with trigonometry and you could do it with coordinate geometry, but it would be incredibly ugly. I've tried that just as an exercise, and it easily fills a full page with mathematical symbols and it leaves no insight whatsoever.

Let's use this idea of focused invariants. What are the natural points, and what are the invariant parts of the figure? Starting with that point A, what kinds of transformations leave things alone, leave parts of the triangle alone? An obvious idea would be to think about rotation about A by 60 degrees because it is an equilateral triangle. That would take the point C to the point B. In other words, the image of C would be B. Let's do this rotation, but let's have fun. Let's let the line $L3$ go along for the ride. Imagine the point C moves into the point B, but as it does, it takes line $L3$ along with it. The image of the line $L3$ I now call $L4$; $L4$ is the image of $L3$. Remember that C was on $L3$, and the rotation took C to B. But B is on $L2$, so the intersection of $L4$ with $L2$ must be where B is. So we located B.

Once we know where B is, the line from A to B is the base of our equilateral triangle, and automatically when we draw an equilateral triangle with that base, the third point C will be on $L3$. It was a magical solution. What happened was truly amazing. By using parts of the figure, just one line and one point and the knowledge that the rotation moved one point of the unknown triangle to another unknown point, we were able to construct the location of both points. We won't always be this lucky to magically resolve a problem with a single rotation.

Now let's study what happens when we compose 2 rotations. The answer is a little surprising. Here's an example of a rotation about point A and a rotation about point B, where we rotate A around A by 30 degrees and B by 15 degrees. If we were rotating around the same center, we know that when you compose 2 rotations, you're just adding the angles. If I just rotate around the center around me by 5 degrees and then by 10 degrees, the net will be 15 degrees. That's not interesting, but what if its 2 centers are different, like in this problem here? Point P gets moved to P' when it rotates around A, and then it gets moved to point P'' when you rotate it about B. It does something, but what will it do? One conjecture is maybe it also rotates by the sum, in this case, by 45 degrees. In fact, that will always be the case.

Here's an example why. If we start with our centers at A and B, so join A and B with a black line, and let's keep track of that black line; it's the only line worth thinking about because we're given 2 points, A and B. When we do our first rotation by A, it'll turn into a red line and the angle, that red angle, is the amount of rotation at A. We know that from the rotated line lemma. When we do the rotation about B, it'll turn into a green line, and that green angle is the amount of rotation at B. We have this picture with a red angle and a green angle, but that green line makes a blue angle with the black line. What's the relationship between these colors? If you look at it carefully, that blue angle is just the exterior angle of a triangle. And a very simple theorem of elementary geometry is that the blue angle will be the sum of the red and the green. That's a simple demonstration. When you do 2 rotations, no matter where the centers are, the angles add. It'll be a brand new rotation with a new center. We don't know where that center is. There are ways of finding it, but that's not relevant for us. The important thing is that the angles add.

There's one case that we didn't think about. What if the angles added up to 360 degrees? For example, suppose the first rotation about *A* was 30 degrees and the second rotation about *B* was 330 degrees. A way to think about 330 degrees is instead of saying that it's 330 degrees counterclockwise, just say it's 30 degrees clockwise. What's happening if they add up to 360 is that one of the rotations is some angle counterclockwise and the other is exactly the same angle but clockwise. If you look at the picture again, we start with our black line, and it rotates into a red line with a red angle. Then its new image is a green line, which rotates the other way with a green angle that's the same magnitude as that red angle. The situation you get is parallel lines. What's happening? What is the net result of this? The image of the black line was that green line; it's just a parallel transport. Well, what does parallel transport? Translation. The key result, this is the lemma we need to remember, is that if you compose 2 rotations and the angles add up to 360 degrees, then the net result, the composition of those 2 rotations, is a translation, possibly the identity translation if the 2 centers are the same. This is an unexpected fact, and it's a very helpful tool, which we'll need for the pentagon problem. Let's call it the reflection-translation tool.

Now, let's get back to the pentagon problem. Remember the problem: We're given the midpoints of the mystery pentagon. What to do? Well, wishful thinking, just like the triangle problem, so let's pretend we know the pentagon's vertices. Draw the picture so that you have something to contemplate. Are there any transformations that you know anything about? Well, the only thing you really know about is the midpoints, so let's start with them. For example, for the midpoint *X*, between *A* and *B*, can we do any rotations that are interesting? Well, if we did a 180-degree rotation, that would just sort of spin around, and *A* would go to *B*, and *B* would go to *A*. Why don't we do that? Let's do that rotation around *X*, 180 degrees, and it will spin *A* into *B*. Then what should we do? Let's just keep going. Spin around *Y* now, and it will take *A*, which is now at *B*, to *C*. Then spin around *Z*, and *A* is going to travel all the way to *D*. Then spin around, and *A* will get to *E*. Spin one more time, and *A* will get back to where it was. That's 5 spins. What's 5 × 180? That's 900. Why don't we do one more spin, a sixth spin? If we do a sixth spin, it will take *A* to *B*, to *C*, to *D*, to *E*, to *A*, to *B* again. In other words, the net travel would be from *A* to *B*. What is 6 × 180? If you add all that up, you get 1080, which is a multiple of 360, so it has to be a translation.

In other words, this 6-rotation composition is the vector going from *A* to *B*. Problem solved, except we don't know where *A* is. Darn. Well, what do we do? We do something. Just pick a random point *P*, and then perform this rotation. *P* spins around *X* to *P*. It goes from *P*, to *P*1, to *P*2, to *P*3, to *P*4, to *P*5, to *P*6. When the dust settles, we have a vector. The vector from *P* to P6 is the exact translation vector from *A* to *B*. That allows us to reconstruct the pentagon because we now know the exact vector from *A* to *B*, and we know the location of the midpoint of *AB*. It's a simple amount of really simple geometric construction to take that midpoint and move the vector on it, and we now can completely reconstruct that side. Once we do so, we can reconstruct the rest of the pentagon because once we have the length *AB*, we can just spin around *Y* and turn *B* into *C*, and turn *C* into *D*, and [turn] *D* into *E*. Before you know it, you've reconstructed the pentagon.

Now what did we do? We did magic here, practically. We used this targeted transformation principle, the idea that we can investigate transformations algebraically, and we came up with a lemma that tells us that certain combinations of rotations unexpectedly turn into translations. We used that fact in a targeted way to take a problem where there seemed to be virtually no information and we were able to reconstruct the unknowns of that problem because we were able to find just one vector. What we glimpsed, just a small amount, is a wonderful world where knowledge of how basic transformations interact with one another allows you to solve these seemingly intractable problems, problems that just don't have enough information. The higher-level reason for this is that we don't use those transformations randomly; we search in a very focused way for transformations that leave parts of our problem invariant. With the pentagon, we didn't do any old rotation; we did 180-degree rotations. By using this sort of focused look for invariance, we're able to decrease the disorder in our problem. We've used this in other contexts before; this is where the imposed symmetry strategy comes from. At an even higher level, the whole entrée to this subject, the idea of doing algebra with transformations, comes from the strategy of flipping one's point of view from the objects of geometry to the motions that leave these objects fixed. This can be done in lots of branches of mathematics. For example, with a polynomial, you could look at the roots of the polynomial that you plug into the polynomial to leave it invariant at 0. You could look at algebraic operations that maybe move these roots around but basically just permute them. For example, if you go way back to our symmetry lecture, we looked at a polynomial of fourth degree, and we did an algebraic substitution, $y = x + 1/x$. That turned out to just permute the roots. It rearranged the solutions of that polynomial.

This idea, which is really an algebraic version of what we've done with geometry, was Galois' inspiration in the early 1800s for a new algebraic approach to the theory of polynomials. It was one of the greatest achievements in all of mathematics. It's called Galois theory, and I'll talk about it very briefly in a few lectures. The point is the giant strategies of flipping the point of view allow you to investigate at higher, and higher, and higher levels. We can do this with algebra. We did it with geometry. We can do it with other kinds of geometries. If we do it with non-Euclidean geometries, this same transformational approach will give you unexpected insights about number theory. But that's a whole other story that we won't get to in this course.

Lecture Nineteen

Building from Simple to Complex with Induction

Scope:

Mathematical induction is the natural way to prove assertions that are recursive, that is, where simpler cases evolve into more complex cases that depend on the earlier cases. Our cornerstone problem is a folkloric tiling of a punctured chessboard, and we also apply induction to combinatorial geometry and a probability problem from the Putnam exam.

Outline

I. Mathematical induction is a proof method closely related to algorithmic proof. It especially works with problems involving evolving structures that build upon simpler structures.

II. Let's begin with a folklore problem that we will eventually solve using mathematical induction: Consider a $2^{2009} \times 2^{2009}$ chessboard with a single 1×1 square removed. Show that no matter where the small square is removed, it is possible to tile this "punctured" chessboard with L-trominos (2×2 squares with one 1×1 square removed).

- **A.** We know that 2009 is a red herring, so we focus on whether it will work for any chessboard of size $2^n \times 2^n$.
- **B.** It certainly works when $n = 1$ or $n = 2$.
- **C.** Can we bootstrap from $n = 2$ to $n = 3$ to $n = 4$ and so on?

III. Mathematical induction allows us to prove an empirical pattern and show how it extends indefinitely.

- **A.** In general, mathematical induction proof involves a sequence of propositions, P_n, indexed by natural numbers. We wish to prove that $P_1, P_2, \ldots$ are all true.
- **B.** To do this, we first show that P_1 (the base case) is true.
- **C.** Then we need to establish the principle that if each case is true, the next one will be as well: In other words, for all positive integers n, if P_n is true, then P_{n+1} will also be true.

D. The assumption that P_n is true is called the inductive hypothesis. We do not know if it is true, but we use its truth to prove that P_{n+1} is true.

IV. A simple example: The plane is divided into regions by straight lines. Show that it is always possible to color the regions with 2 colors so that adjacent regions are never the same color.

A. Let's call colorings such as the above "nice." Define P_n to be "If the plane is divided into regions by n straight lines, then it is possible to color nicely."

B. This is clearly true for $n = 1$ (we have proven the base case).

C. But what about, say, $n = 10$? We need an algorithm for moving from the nth case to the $(n + 1)^{\text{th}}$ case.

D. Keep things concrete: Suppose I can nicely color any 5-line configuration. How can I use this to nicely color an arbitrary 6-line configuration?

E. Here is the formal solution for the inductive step.

1. Suppose P_n is true. Given any $(n + 1)$-line configuration, temporarily ignore one of the lines.
2. Now you have an n-line configuration, which you know you can color nicely.
3. Finally, invert the colors on one side of the $(n + 1)^{\text{th}}$ line.

V. Another problem about lines: Lines in a plane are in general position if no 2 are parallel and no 3 meet in a point. If 10 lines are drawn in general position in the plane, into how many regions do they divide the plane?

A. Clearly we want to discover a formula for n lines.

B. If we let R_n denote the number of regions made by n lines in general position, we conjecture that $R_n = n + R_{n-1}$.

C. Notice that this is the key to a rigorous induction proof, because it actually suggests the way that you go from the $(n - 1)^{\text{th}}$ case to the nth case (from one case to the next case).

1. Suppose we have $n - 1$ lines in general position, creating R_{n-1} regions.
2. Imagine drawing a new line so that it is not parallel to any of the other lines and does not intersect them in any of the previous intersection points.

3. This new line will intersect each of the $(n - 1)$ old lines, producing a new region each time an intersection is achieved.
4. When the new line intersects the last of the old lines and exits, one final new region will be produced, for a total of n new regions.

VI. Here is a probability example from the 2002 Putnam exam: Shanille O'Keal shoots free throws on a basketball court. She hits the first and misses the second; thereafter, the probability that she hits the next shot is equal to the proportion of shots she has hit so far. What is the probability she hits exactly 50 of her first 100 shots?

A. After the second toss, the proportion of successes is 1/2, so on toss 3, we have a 1/2 chance of getting another basket.

B. For toss 4, it gets more complex. We employ the draw a picture strategy loosely and create a tree diagram that shows all the scenarios.

C. The outcomes have equal probability! This is somewhat surprising, but we can attempt to prove it with induction.

D. Let $P(b, t)$ denote the probability that we get b baskets in t tosses. Our conjecture is that $P(b, t) = 1/(t - 1)$ for each of the valid values of b (between 1 and $t - 1$).

E. We will prove this by induction on t. This time, our base case is $t = 2$, and it is trivially true.

F. Suppose that $P(b, t) = 1/(t - 1)$ for some value of t greater than or equal to the base case of 2. We will use this to prove that $P(b, t + 1) = 1/t$.

G. How do we get b baskets in $t + 1$ tosses? Either we get a basket on the $t + 1$ toss, or we do not.

1. Suppose we do not get a basket. Then we accumulated b baskets in the first t tosses. By the inductive hypothesis, this has probability $1/(t - 1)$. At this point, the probability we will get a new basket is equal to the current proportion, b/t. So the probability that we end up with b baskets by missing the last toss is $[1/(t - 1)](1 - b/t) = (t - b)/t(t - 1)$.

2. Now suppose we do get a basket on the final toss. Then we accumulated $b-1$ baskets in the first t tosses. By the inductive hypothesis, this also has probability $1/(t-1)$. The probability we will get a new basket is equal to the current proportion, $(b-1)/t$. So the probability that we end up with b baskets by missing the last toss is $[1/(t-1)](b-1)/t = (b-1)/t(t-1)$.
3. Adding these 2 probabilities, we get $(b-1)/t(t-1) + (t-b)/t(t-1) = (t-1)/t(t-1) = 1/t$.

H. The induction proof is formally correct but not fully illuminating. This is a feature of some induction proofs. You can verify how to get from t to $t+1$, but you sometimes do not know why it works.

VII. Let's return to the tromino problem. Here is a way to go from $n=3$ to $n=4$.

A. The crux idea: symmetry!

B. Given an arbitrary 16×16 tile board missing 1 tile, without loss of generality, the hole is in the southwest quandrant.

C. Place a tromino in the center so that it takes a single bite out of each of the other quadrants.

D. Now all 4 quadrants have a single hole; by the inductive hypothesis (for the 8×8 minus 1 board), we can tile each of them!

E. Clearly this can be generalized; this is our inductive algorithm.

Suggested Reading:

Fomin and Itenberg, *Mathematical Circles*, chap. 9.

Goodaire and Parmenter, *Discrete Mathematics with Graph Theory*, sec. 5.1.

Maurer and Ralston, *Discrete Algorithmic Mathematics*, chap. 2.

Problems:

1. Conjecture a formula for the sum of the first n Fibonacci numbers. Then prove your formula by induction.
2. Prove that $F_n < 2^n$, where F_n denotes the nth Fibonacci number.

Lecture Nineteen—Transcript

Building from Simple to Complex with Induction

Back in Lecture Five, when we learned about closing the deal, we spent some time learning basic proof techniques. Direct algebraic proofs and proof by contradiction have become our basic tools for demonstrating the truth of a mathematical proposition, in other words, for proving things. There are other ways to prove things. We have seen the use of algorithmic proofs, where we create a sort of machine that solves our problem, proving that it can be done, which validates a get your hands dirty strategy. Mathematical induction is another proof method that's closely related to this, but first, how do we know what proof methods work for what? When I'm teaching problem solving to my students, I'll often put a problem down on the blackboard and I'll ask them, "The problem is screaming at you for a proof. What method is the problem screaming that you use?" If you just sit back and look at the problem and say, well, is the problem asserting something that's really complicated, such as infinity? Then proof by contradiction is often a good approach because then you begin by assuming finiteness. You have easy things to work with. Induction has its own context. Inductive proofs usually involve problems that involve evolving structures that build upon simpler structures or problems that have recurrence, for example, the recurrence relation defining the Fibonacci numbers, that the nth Fibonacci number is equal to the sum of the $n-1^{\text{st}}$ and $n-2^{\text{nd}}$. Problems where we have recurrence or where we think there might be recurrence, this is the natural arena for mathematical induction. Sometimes the structural process, the evolving process, cannot be easily captured just with an algebraic equation, though.

Let's look at a fun folklore problem that we'll solve using mathematical induction. It's called the tromino problem. A tromino is one of the many polyominoes. A tromino has 3 parts to it. It's just a tiny L. Imagine a 2×2 square with a little corner bitten out of it. It's a 2×2 L shape. That's what a tromino is. Consider a gigantic chessboard $2^{2009} \times 2^{2009}$, and we remove a single square from this gigantic chessboard. Show that no matter where that tiny square is removed, it's possible to tile this giant square minus the tiny square with these trominoes. Sometimes they're called ells or ell-trominoes. If this is true, then that tells you that $2^{2009} \times 2^{2009} - 1$ would have to be a multiple of 3 since each of the little ells has 3 squares in it. We know from our basic problem solving that 2009 is a red herring, so let's focus on whether it will work for any chessboard of size, not 2^{2009}, but 2 to some power times 2 to some power; let's try $n = 1$, $2^1 \times 2^1$. It works for a 2×2 easily because that's just a single tromino. How about changing n, our exponent, to 2? Now we have $2^2 \times 2^2$, in other words, 4×4, with one thing removed, and if we just try one example, you can see it works, at least by trial and error. If we move that little hole around, we'd find that no matter how we move the hole around, we can do our perfect tiling, so no overlapping and we cover every square. The question is: Can we bootstrap, sort of bring ourselves up from $n = 2$ to $n = 3$ to $n = 4$, so powers of 2^2, 2^3, and 2^4, and so on? That's why the problem is screaming at us to use induction, this kind of a structure.

How do we do it? What is induction? Mathematical induction is only useful if you understand an empirical pattern and can show how it extends indefinitely. Mathematical induction is sort of the rigorous form of ordinary induction in the everyday sense, where we're using sort of the premise that what you've already discovered empirically tells you what will happen in the future. That previous experience will inform future experience. That's just everyday life. Mathematical induction is a rigorous technique that allows you to prove the next stage based on the previous stage. What it allows you to do, if it's done properly, is to prove infinitely many statements in one fell swoop, actually 2 fell swoops. Here's the idea: Suppose you have infinitely many dominoes labeled 1, 2, 3, and so on, and you know only 2 things about them. Number 1: You know that the first domino will fall towards the right, fall towards the next domino. Principle number 2 is that whenever a domino falls, it will cause its next-door neighbor also to fall. In other words, if domino n falls, then so will fall domino $(n + 1)$. Clearly, all the dominoes will fall. In general, that's how mathematical induction proofs work. We use a sequence of propositions, P_1, P_2, P_3, where these are all sentences. P_1 is the first proposition, P_2 is the second proposition, and so on. What we want to do is prove that all of these propositions are true. For the tromino problem, P_1 is the proposition about tiling a $2^1 \times 2^1$ chessboard with the hole removed. P_{17} is a $2^{17} \times 2^{17}$ with the hole removed. We want to prove P_{2009}, but what we'll do instead is prove all of them, P_1, P_2, the infinite sequence. To do this, all you need to do is show that P_1 is true. That's called the base case; that's knocking over the first domino. Once we've done that, we now need to establish the principle that if any domino falls, it will knock over its neighbor to the right. That's the inductive step. In other words, what we want to prove, using more algebraic language, is if n is an arbitrary positive integer, that if we knew that P_n was true, then we would know that $P_{(n+1)}$ is also true. The assumption that P_n is true is called the inductive hypothesis. Note that we don't know if it's true, but we just assume it's truth to prove that $P_{(n+1)}$ is true. It's a kind of strange process where we just assume the truth of something, but that's well known to us problem solvers. All we're using is wishful thinking. We go back to the fact that we have already, hopefully, proven the base case. If P_1 is true, then the inductive step allows you to turn P_1 into P_2, and P_2 into P_3, and so on. Using that sort of fantasized inductive step plus the truth of the first step will give you all of the steps. That's the structure.

Let's recap. If you've never done this before, it's kind of an unusual process. Mathematical induction requires 2 parts—the base case, usually easy, a proof that the first statement in your sequence is true, and then the inductive step, which begins by fantasizing that the nth step is true, where n is an arbitrary positive integer, and we use that to logically show that this will imply the truth of the $n + 1^{\text{st}}$ statement. Let's try a simple example. Imagine the plane is divided into regions by straight lines. I don't know about you, but when I was a kid, I remember one time this friend of my older stepbrother wanted me to get out of his hair. He said, "Hey, I have something fun for you to do." He took a piece of paper, and he just drew 50 lines on it. He said try coloring it like a chessboard. I spent about 20 minutes doing that, and I wasn't bothering these older teenagers. Here's the mathematical question: Given a bunch of lines on a piece of paper, is it always possible to color the regions like a chessboard, in other words, so that adjacent regions are never the same color? What does this have to do with induction? It's a problem about finitely many lines, and clearly, it's an assertion about any number of lines. It's also true that if we draw 50 lines, then we've already done the subset of the 51-line configuration. It's the proper venue for induction; it's something where we have this increasing structure.

What is P_n? P_n would be the proposition that if the plane is divided into regions by n straight lines, then it's possible to color them in this—let's call it—this nice way. We'll use the abbreviation nice for this chessboard coloring. Clearly, when you have 1 line, the base case is true. You just draw a line and color one side black, one side white. You're done. And $n = 2$ is obvious, and $n = 3$ is pretty obvious. What about $n = 10$? If you have 10 lines, it's not obvious that you can color this like a chessboard. You might do it and be successful, but there are so many possibilities maybe you just got lucky. How do you prove it rigorously? The idea is to figure out the recurrence. Figure out how you get from, say, 9 lines to 10 lines. A good way to focus your mind here is to stay away from algebraic symbols, like n going to $n + 1$. Instead, keep it concrete. Remember, back in an earlier lecture, we mentioned the sad fact that human beings are basically pretty stupid. That's not a pejorative here; it's just the truth. The truth is it's a lot easier to think about the number 5 than the algebraic symbol n. Here's the question: Suppose I can nicely color any 5-line configuration in this chessboard way. Could I use this to nicely color an arbitrary 6-line configuration? Remember, I'm just fantasizing about my success with 5, but if I could do 5, could I do 6? In other words, could we extract the 5-line situation from any arbitrary 6-line configuration? That's easy.

Here's an illustration that shows you how you can get from 2 lines to 3 lines using metaphorical symmetry. We start with a simple chessboard with 2 lines, a nice, simple pattern, and then we have that third line in there. Notice that as we drew that third line, we just inverted things, inverted things to the right of the line. We could do that in general. Here's sort of a formal solution. Here's how the inductive step would go: Suppose P_n is true, so that means you can nicely color any n-line configuration. Now, suppose someone just plops an $(n + 1)$-line configuration in your lap. Temporarily ignore one of the lines; just remove it for just a second. Now what you have is an n-line configuration, which you know you can color nicely. That's your fantasy, remember, so color it nicely. Then, that line you were ignoring, it's still there. Just invert the colors on one side of that line. Voilà, we clearly have created a new nice coloration of the $(n + 1)$-line configuration. You also might remember that it seems a little reminiscent to that old social engineering problem because it's using this same idea of sort of a flipping, of a metaphorical symmetry flip of 2 states.

Let's do one more lines problem. Imagine that you have lines in the plane in what's called general position. That means no 2 are parallel, so you don't have lines like this, and no 3 meet in a point, so you won't have 3 lines meeting together. Suppose you drew 10 such lines in general position. How many regions would they divide the plane?

You know that's a pretty ugly thing to try to draw, and unless you had a ruler, there's no way you'd do it properly. Counting the regions would drive you crazy. This problem is secretly asking you to do an inductive investigation. It's saying start with 1 line, 2 lines, 3 lines, and build up the structure into 10 lines, and along the way, figure out how you're going from one stage to the next. That's the real idea of induction. Remember, we're not in the business of doing proofs in this course; we're in the business of investigating mathematical problems. What we're looking for here are problems where we have evolving structures. Induction is just a logical way to help us explore that. It's the investigation that matters the most. Let's investigate; let's get our hand dirty.

With 2 lines, clearly, there are 4 regions. With 3 lines, we can count, and it looks to me that there are about 7 regions. There are 4 lines, 5 lines; let's keep track of our results here. When there was 1 line, we had 2 regions; 2 lines, 4 regions. We're about to think of powers of 2, but 3 lines gave us 7, 4 lines gave us 11, and 5 lines gave us 16 regions. At this stage of the investigation, we most likely do not understand what we're doing, but we've gathered data. It's quite valid now to look for a pattern, and it's easy to see that there's a pattern in there. There's an obvious conjecture because if you go from 2 to 4, at first you were doubling, but now you know doubling doesn't work. So look at something else. The simplest thing is 2 to 4, add 2; 4 to 7, add 3; 7 to 11, add 4; 11 to 16, add 5. Looks like a pattern. The question is: Is it a robust pattern?

Let's use a little bit of notation. We shall let our R_n denote the number of regions made by n lines in general position. What we're conjecturing, then, is a recurrence relation. We're conjecturing that R_n, the number of regions for n lines, is equal to $R_{(n-1)} + n$. If it's true, this is an easy key to a rigorous induction proof because it actually suggests precisely how you go from the $n - 1$ case to the nth case. That's the point. If you can discover a recurrence, you're done with an induction, and in fact, most of the time, when somebody is writing a mathematical paper or giving a talk, they will establish this recurrence and say, therefore, it's true by induction. That's the entire proof. They don't go into the details because the heart of it is the recurrence. Let's try to prove this recurrence. We've observed it just with numbers, but now, we have to go back and look at our picture. What we want to figure out is how you turn an $(n - 1)$-line configuration into an n-line configuration and keep track of the area.

Suppose you have $(n - 1)$ lines in general position, and they're creating, presumably, $(R_n) - 1$ regions. Now imagine drawing in a new line, also in general position. It's not parallel to any of the other lines, and it does not intersect any of them at any of the previous intersection points. We'll illustrate this by adding a new line in red. Imagine it coming in dynamically. Actually, imagine it coming from left to right, sort of just zooming in to our previous configuration. Because of general position, this new line will intersect each of the $(n - 1)$ old lines. It has to because it's not parallel to any of them. Each time it intersects a new line, it will produce a new region. Every time you hit an intersection point, a new region is born. That's going to give you $(n - 1)$ new regions, but when that new line intersects the very last of the old lines and then exits out to infinity, it will create one final new region for a total, then, of n new regions. We've proven the recurrence. The reason we've proven the entire fact is because we know the base case is true. We know that R_1 is equal to 2, and then the next one is equal to $(2 + 2)$, and then $+ 3$, and then $+ 4$, and so on. What we could do is work out a formula because we know how to add sums of integers. In general, what we've deduced is that (R_n) is really equal to 1, the very beginning part, $+ 1 + 2 + 3$ up to n. But it's not really necessary to write a formula. Really, the core is the recurrence that $(R_n) = (R_{n-1}) + n$. From that recurrence, we know everything about the sequence of numbers, and remember, the original question was to find R_{10}, and we could easily compute that. That actual value is not what's important. What's crucial here is understanding the notion of recurrence and discovering the proper recurrence.

Here's an example from the Putnam exam. Remember, the Putnam is this incredibly hard undergraduate problem-solving competition, where the median score is generally, most years, 0 out of 120. If the Putnam is easy, I think last year it was easy, the median score was 1 out of 120 points. Here's an example of what the Putnam considers one of its easiest questions. It was the first question of the 3-hour afternoon session of 6 problems in 2002. Shanille O'Keal shoots free throws on a basketball court. She hits the first and misses the second, and thereafter, the probability that she hits the next shot is equal to the proportion of shots she has hit so far. What is the probability she hits exactly 50 of her first 100 shots?

By now we're sophisticated, and we know how to make it easier and get our hands dirty with smaller examples. After the second toss, the proportion of successes is 1/2, so on toss number 3, we have a 1/2 chance of getting another basket. The 2 outcomes of one or 2 baskets each will have probability of 1/2. For toss number 4, we need to work a little bit harder. What we'll employ here is the simplest draw a picture strategy because we need to organize our work in some way. It's a probability question, so the natural thing to try is a little probability tree diagram that shows all of the scenarios. For example, if we look at 4 tosses, when we go up on the tree, we're missing, and when we go down, we're getting a shot. We get our first shot, and we now have one basket, and then we have a 50/50 chance of getting another basket. For example, if we go down to get 2 baskets, that was a 1/2 probability event, and now the probability is going to be 1/3 that won't get a basket or 2/3 that we will. When we work out all of the probabilities, you discover that at each stage, for the outcomes, they're all the same. This is kind of unexpected, but the outcomes are the same probability. This is somewhat surprising, but we can attempt to prove it with induction because this is a natural case where we can build the t toss situation from the $(t - 1)$ toss situation.

Let's use some notation. Let's let P for probability, and we'll give it 2 variables, b and t, so $P(b, t)$ is the probability that we get b baskets in t tosses. Our conjecture, then, is that the probability of getting b baskets in t tosses is going to be $1/t - 1$ for each of the valid values of b that lie between 1 and $t - 1$. For example, with 4 tosses the probabilities are all the same, namely, 1/3. We will prove this by induction on t, the number of tosses. For example, imagine 50 tosses. We'll first figure out what happens with 49 tosses, and then we'll just build from 49 to 50. What's good about this is if you can fantasize that you know everything about, say, the 50th iteration of this incredibly complicated probability tree, we don't have to worry about the entire tree. We just need to take a single node in the tree, and then there are only 2 things to get to the next level of the tree. What's our base case? Our base case has to start with $t = 2$ because the problem doesn't even begin until the second toss. And then it is trivially true. There's only one outcome. Now for the inductive step: Suppose we fantasized that $P(b, t)$, the probability of b baskets in t tosses, is equal to $1/(t - 1)$, so it's independent of b, for some value of t greater than or equal to that base case of 2. What we now want to do is move up to the next level and prove that the probability of getting b baskets in $t + 1$ tosses is $1/t$. How do we get b baskets in $t + 1$ tosses? Either we get a basket on the $t + 1$ toss or we don't.

Suppose we don't get a basket. Then, we accumulated b baskets in the first t tosses. By the inductive hypothesis, this has probability $1/(t-1)$. Remember, this is complete fantasy; we're just fantasizing that we know this. That's the beauty of induction. At this point, the probability that we'll get a new basket is equal to the current proportion, which is b/t. The probability that we end up with b baskets, by missing the last toss, is equal to $(1/(t-1))$—that's our inductive hypothesis—$\times$ $(1 - b/t)$, which simplifies into $(t-b)/(t(t-1))$.

Suppose we do get a basket on the final toss. Then we accumulated $b-1$ baskets in the first t tosses. By the inductive hypothesis, this also has probability $1/(t-1)$. Then the probability that we will get a new basket is equal to the current proportion, which is $(b-1)/t$. The probability that we end up with b baskets in this scenario, by missing the last toss, is equal to $(1/(t-1))$, as before, $\times$ $(b-1)/t$, which is $(b-1)/(t(t-1))$. Adding these 2 probabilities, we get $(b-1)/(t(t-1)) + (t-b)$ over the same denominator, which reduces to $(t-1)/(t(t-1))$, and when you cancel, you're left with $1/t$. It worked almost like magic. In fact, that's the problem. The induction proof was formally correct and algebraically impeccable—at least, I think algebraically impeccable—but it was not completely illuminating. In fact, it was really not illuminating at all. It was something a machine could have done. This is, unfortunately, a feature of some induction proofs. You're able to verify how to get from t to $t+1$, but you sometimes are not sure why it works. This issue of how versus why is very, very important, and we will address it later, in our final lecture.

Let's end our lecture with an induction that's fun. I did the Shanille O'Keal problem, not to bore you or to disgust you with a lot of algebra, but because that's often how induction has to be done. You realize you have a structure that's building from one level to the next, and sometimes the only method of investigation is an algebraic one. With our tromino tiling problem, we can have more fun. Let's think of an algorithm for tiling an arbitrary $2^{n+1} \times 2^{n+1}$ chessboard minus a single tiny square that uses, in some way, the fact that we're fantasizing that we can tile a $2^n \times 2^{n-1}$ board. How can we do that? For example, let's make it concrete. Let's take an arbitrary 16×16 board where we take 1 little square out. It's a 16×16 with a hole. What we want to do is figure out if we can relate it to an 8×8 case, and we're assuming we know how to do an 8×8 case. Without loss of generality, the hole has to be somewhere. Let's assume it's in the southwest quadrant. Let's use some symmetry. We have a hole in the southwest, so the southwest can be tiled by the inductive hypothesis. If only we had a hole in the northeast, or a hole in the southeast, or a hole in the northwest. Can we put holes there? Sure, we can. What we need to do—and here we're really being inspired by symmetry—is put your tromino in the center of the board oriented so that the ell eats up a little corner of the northwest, a little corner of the northeast, and a little corner of the southeast. Each of those 3 quadrants has a hole chopped out of them, and of course, we were assuming that there was a hole somewhere in the southwest anyway. Now by the inductive hypothesis, each quadrant can be tiled on its own. We have done our induction. We've proven the induction because we were able to use 8 in order to prove 16. We can use 16 to prove 32, and then prove 64, and so on. We've proven it for $2^{2009} \times 2^{2009}$, but that really wasn't the point. What was important was we established the recurrence, and here, establishing the recurrence was fun. It wasn't algebraic, but it used a beautiful construction.

Again, let's get back to the question: Why use mathematical induction? Sometimes you have to. Remember, we know why proof by contradiction works. If you want to prove that something is infinite, starting with the negation of it, that it's finite, gives you something easier to analyze. In this course, the high-level view, looking for the why, is very important. Mathematical induction, we now know why it works if we have a recursive structure where our problem is building, one structure getting larger and larger and larger. Mathematical induction is not nearly as applicable as proof by contradiction. That's why we didn't study it earlier, but it is pretty much the only way to look at situations that have this recursive property. The real issue is that sometimes asking the question "Is this a recursive problem?" is half the battle. When you look at a problem, you want to ask yourself: Is this a problem where we're building a complicated structure from a simpler one? If that's the case, then induction is probably the right way to investigate it. This question—is it recursive?—is a question that we will ask in later lectures with very dramatic results.

Lecture Twenty
Induction on a Grand Scale

Scope:

What is the probability that a randomly chosen number in Pascal's triangle is even? This problem is surprisingly easy to investigate but requires sophistication to resolve. By this stage, you have a good grasp of investigative methods, summation, mathematical induction, and modular arithmetic, so you are ready for this investigation, the first of the advanced lectures as we approach the end of the course.

Outline

I. What it the probability that a randomly chosen member of Pascal's triangle is even?

A. This is a meaningless question as posed; Pascal's triangle is infinite!

B. Reformulation: Let P_n be the probability that a randomly chosen element from the first n rows of Pascal's triangle is even. Does P_n converge to something as n approaches infinity?

C. We are asking a question about density. Define the density of a subset S of the natural numbers by computing the probability that a randomly chosen integer from the first n integers is in S. Then we see what happens to P_n as n gets arbitrarily large. If it converges, then that is S's density.

1. The density of the even integers is 1/2.
2. The density of the perfect squares is 0, since among the first n^2 integers, exactly n are squares, so the relative frequency is $1/n$, which gets arbitrarily small.

II. Recall that we called the elements of Pascal's triangle binomial coefficients and asserted that the elements of row n Pascal's triangle were the coefficients of the binomial $(1+x)^n$.

A. For example, $(1+x)^4 = x^4 + 4x^3 + 6x^2 + 4x + 1$, and indeed, the coefficients are the numbers in row 4.

B. This is actually quite easy to prove, now that we know about induction. Let P_n be the statement that the coefficients of $(1+x)^n$ are the elements of row *n*.

C. The base case is obvious, since row 0 is just 1, and row 1 is 1, 1.

D. We want to prove the inductive step now, but let's keep it concrete and informal. We will just show that P_5 implies P_6; our argument will generalize easily.

E. Start with the inductive hypothesis: $(1+x)^5 = 1x^5 + 5x^4 + 10x^3 + 10x^2 + 5x + 1$.

F. Thus $(1+x)^6$ will be this polynomial, multiplied by $(1 + x)$.

G. This is exactly how we get row 6 of Pascal's triangle: $(1+x)^6 = 1x^6 + 6x^5 + 15x^4 + 20x^3 + 15x^2 + 6x + 1$.

III. Now let's look at the parity of the numbers of Pascal's triangle. We will work (mod 2).

A. Look at the first 9 rows. At first, there are not many evens at all. But row 4 and row 8 are all even, except for the ubiquitous 1s that start and end every row. And notice that rows 3 and 7 are all 1s.

1. We conjecture that row will be all 1s and that row 2^n will be all 0s, except for the first and last terms.

2. Clearly, the second statement follows from the first, but how do we prove the first statement?

B. Now look at rows 0–32.

1. We see a fractal structure, with inverted triangles of 0s. What causes them?

2. The seed of the 0 triangles is the row of all 1s, since this forces the next row to be all 0s (except the first and last terms).

3. The natural way to look at the parity of Pascal's triangle is by successive doublings of it.

C. Let T_n be the nth-order triangle that ends with a row of 1s. T_1 is the triangle consisting of rows 0 and 1 (i.e., it contains three 1s), and T_2 is the triangle that is built out of 3 copies of T_1, with a 0 in the middle.

D. In general, T_n is the triangle that ends with row $2^n - 1$, which is all 1s. This starts 2 seeds at opposite ends, with 0s in between, which then grow 2 more copies of T_n, producing a new structure, T_{n+1}.

IV. Now that we have inductively proven the fractal structure of Pascal's triangle, we can try to count the even terms. This turns out to be complicated, but using the flip your point of view strategy, we instead look at odd terms.

A. This is nearly trivial, since T_1 has exactly three 1s, and T_{n+1} is composed of 3 copies of T_n with 0s in the center. If we define U_n to be the number of 1s in T_n, we get the simple formula $U_n = 3^n$.

B. Our final step is to compute the relative fraction of 1s in T_n and then let n get large.

1. How many elements are in T_n? It is a triangle starting with one element and ending with 2^n elements. So the number of elements in T_n is the 2^n triangular number!
2. This is equal to $1 + 2 + \cdots + 2^n = 2^n(2^n + 1)/2$.

C. Thus, the probability that an element in the first 2^n rows is odd is equal to $2(3^n)/2^n(2^n + 1)$.

1. Despite the factor of 2 in the numerator, the 4^n in the denominator will eventually overpower it, so the limit is 0.
2. A more rigorous way to see this is by dividing numerator and denominator by 4^n.
3. As n grows larger, the entire fraction approaches 0.

D. So the probability that an element is odd approaches 0. That means that the probability an element is even approaches 100%, which is truly surprising. In other words, essentially all binomial coefficients are even!

V. We just proved an absolutely amazing fact about long-term convergence of parity, an asymptotic property of Pascal's triangle. But it would be nice to analyze the parity in a more exact way. In Lecture Four, we counted the number of evens and odds in each row, and the number of odd terms was a power of 2. Which power of 2?

A. What is the appropriate point of view for investigating powers of 2? The binary (base-2) system, where we write numbers as sums of powers of 2.

B. When we make a table of the number of odd terms in each row and look at the row numbers in binary, the conjecture is clear: The number of 1s in row n is equal to 2 raised to the number of 1s in n when n is written in base 2!

VI. But why? Remember that the elements of Pascal's triangle are the coefficients of $(1+x)^n$. When $n = 2$, we have $(1+x)^2 = 1+2x+x^2$, but (mod 2), the middle term disappears.

A. So $(1+x)^2 = 1+x^2$ (mod 2). If we square this again, we get $(1+x)^4 = (1+x^2)^2 = 1+x^4$ (mod 2).

B. In general, we see that for any n, $(1+x)^{2^n} = 1+x^{2^n}$ (mod 2).

C. This immediately explains why row 2^n has just two 1s! But what about an arbitrary row, say, row 11?

1. Write 11 in binary: 1011.
2. Then look at row 11 of Pascal's triangle by expanding $(1+x)^{11}$.
3. We see that $(1+x)^{11} = (1+x)^8(1+x)^2(1+x)^1$, using the binary representation. There were three 1s in the binary representation for 11, and hence there are 3 terms in the product.
4. But reducing this modulo 2, we get $(1+x)^{11} = (1+x^8)(1+x^2)(1+x)$, a product of 3 binomials. When they are multiplied out, we will have 8 different nonzero terms.

D. Here is a slicker way to see it: When $(1+x)^{11}$ is multiplied out and simplified modulo 2, it will be a sum of powers of x^n, where the coefficients will either be 0 or 1.

Suggested Reading:

Edwards, *Pascal's Arithmetical Triangle*.

Tabachnikov, *Kvant Selecta*, chap. 1.

Problems:

1. There is a fun pattern in Pascal's triangle: Row 0 is 1, row 1 is 11, row 2 is 121, row 3 is 1331, and row 4 is 14641. Notice that for each *k*, row *k* is 11 raised to the kth power! Explain why this pattern is true and why it fails for *k* greater than 4.
2. Investigate the same question that we did in the lecture, but modulo 3. In other words, look at the patterns of when elements of Pascal's triangle are multiples of 3. It is a little more subtle than before, because now there are 3 possible values (mod 3): 0, 1, and 2.

Lecture Twenty—Transcript

Induction on a Grand Scale

This is the first of what will be 4 lectures that I classify as advanced as we approach the end of our course. For example, 3 out of the next 4 lectures, including this one, will investigate infinite structures. Here's a way to understand what I mean by advanced. You could talk about classifying hikes as intermediate, easy, advanced, and you could have an advanced hike, but it might be advanced only because it's a pretty easy hike, but there's one scary section where you kind of have to go around a cliff edge, or maybe it's very steep throughout, or maybe it's steep and it's scary. This first lecture is a relatively easy advanced hike. You can think of this as moderately steep, because we'll be using quite a bit of mathematics, but not too scary. There might be one or 2 scary bits, but nothing that you can't handle. Speaking of handling, at this advanced stage, you might wonder, am I ready? The answer is, yes, you're ready. Definitely use your pause buttons just like you've used before. The problems are hard, but you're always ready to investigate them.

In this particular lecture, we bring together many mathematical techniques that we have already studied, such as induction [and] summation, and we also use new ideas, including convergence of probabilities and the idea of a natural point of view for looking at certain types of numbers. Our mathematical ideas are pretty sophisticated, but the problem-solving ideas are not. We will be richly rewarded just by careful experimentation. The object of our investigation is Pascal's triangle, which we first saw back in Lecture Four when we were just getting started with numerical experimentation and conjectures. I mentioned that Pascal's triangle was a great playground for mathematical investigations, and the reason it is is because it's connected to so many different things. It's not just a sequence of numbers, but it's also connected, as we'll see, with polynomials. We conjectured it, but we'll prove that here. Also, since Lecture Four, we've developed the idea of mathematical induction and the idea of larger and larger structures and recurrences, and we're going to start applying that with Pascal's triangle.

Now we're ready to look at Pascal's triangle again, and we'll investigate a very interesting, deep, unexpected phenomenon. Before we begin, let's review what we know about Pascal's triangle. Remember, it starts with sort of the seed of the number 1, with 2 1s below it, and then you just build up by summation of this sort of funnel principle. The 2 1s give you a 2. There are 1s on either side, and then the next row is 1, 2, 1. Then just adding the successive terms, we'll get 1, 3, 3, 1, then 1 + 3 = 4, 3 + 3 = 6, and so on. The bottom row that you see goes 1, 5, 10, 10, 5, 1. We call that, by the way, row 5 because we start at row 0. That's a very important notational trick. Remember in Lecture Four, we started to investigate parity, and we actually counted the number of even and odds in each row. We noticed that the number of odd terms seemed to be a power of 2, and the question was: Which power of 2? We're not going to look at the row-by-row question yet. Instead we're going to change our perspective and we'll ask sort of a holistic question. We're going to ask: Where's the odd- and evenness in all of Pascal's triangle? In other words, I am going to ask kind of crazy question. What is the probability that a randomly chosen member of Pascal's triangle is even? It's a crazy question because it doesn't make sense because Pascal's triangle is infinite. Here's a better way to think about it. Imagine that you had a lot of time and you wrote out many, many rows of Pascal's triangle, like a couple quadrillion rows. Then you took all of those numbers, and you put them on little pieces of paper, and you put it in a hat. Then you shook the hat really, really well, and you picked out a number at random, and you check to see if it's odd or even. What's the probability that that number will be even? That's the question we're going to ask, but since Pascal's triangle is infinite, we have to think about an infinite process.

Here's a more formal reformulation. We'll let p_n be the probability that a randomly chosen element from the first n rows of Pascal's triangle is even. Again, imagine taking n rows. Chop up all the little pieces. Put it in a hat. Shake it really well, and pick out a piece, and check to see if it's even. This is a probability that, in some sense, could be computed. For example, if there are a billion elements and 500 million of them were even, then the probability would be 1/2. For each n, there's a p_n, and so we have a sequence of numbers, P_1, P_2, P_3, P_4, and so on. The question is: Does this converge to something as n approaches infinity? Convergence is kind of a tricky issue. This is a major topic in calculus, and this is not a calculus course. Your intuition already understands the idea of convergence of what we call relative frequency. For example, suppose you were just flipping a coin, heads and tails. Here's a computer simulation that keeps track of the number of heads versus the numbers of tosses over 6000 tosses. At first, there's no real pattern to it, but after we've done a couple hundred tosses, it settles down. You might have a long run of heads for a while or a long run of tails, but the long-term relative frequency approaches 50%. That, in a sense, is what it means to say that when you flip a coin, the probability is 1/2 that you're going to get heads. You never know what an individual coin is going to do, but you can talk about a long-term frequency. We're going to do the same thing with Pascal's triangle, only it's not random. The numbers are all there. They've all been determined to eternity.

The mathematical notion of relative frequency is a notion called density. For example, what is the density of the even numbers? This is an easy one. What that means is if we were to go out along all the integers and count the evens and compute the long-term relative frequency of evenness as we go further and further out, what does it converge to? For example, go out to the first 10. How many evens do you see? 5. What's the frequency? 5/10. Go out to 11. Now it's 5/11. Then it's 6/12. Then it's 6/13, 7/14, and so on. You can see that these long-term frequencies are going to be either exactly 1/2, or they'll be off by just a tiny, tiny bit, and they will slowly but surely converge to 1/2. Density is the same idea as long-term relative frequency, but it's sort of a more rigorized notion, using formal convergence, which we don't have to worry about. Just use your intuition about this. Here's another density example. The density of the odd numbers would also be 1/2. How about the density of the integers, the positive integers that leave a remainder when you divide them by 7, in other words, the numbers that are congruent to 3 (mod 7)? One out of every 7 numbers is like that. It's a completely rigid pattern starting with 3, then 10, then 17, and so on, so the density would be 1/7. How about the density of the perfect squares? That turns out to be 0 because among the first n^2 integers, exactly n of them are squares. The relative frequency is n/n^2, which is $1/n$, but as n gets large, that gets arbitrarily small. It approaches 0, and we say the density is 0.

Let's go back to Pascal's triangle and see how it does relate to polynomials. Recall that we called the elements of Pascal's triangle binomial coefficients, and we asserted without proof that the elements of row n were indeed the coefficients when you expand the binomial $(1 + x)^n$. For example, if you look at row 5 of Pascal's triangle, you could get those coefficients by taking $(1 + x)^5$. You get $1x^5 + 5x^4 + 10x^3 + 10x^2 + 5x + 1$. That's why Pascal's triangle is so rich for investigation. It's not just sequences, but it has a holistic sense to it. Each row is actually a single thing, in a sense. This is actually very easy to prove now that we know about induction. Let S_n denote the statement that the coefficients of the nth row are the expansion of $(1 + x)^n$, that the elements of row n are those coefficients. The base case is very simple. Row 1 is 1, 1, and it's the coefficients of $(1 + x)$, so we've done the base case.

To prove the inductive step, we could be very formal and show that S_n implies $S_n + 1$, but as I've told you before, it's better to stay concrete. Let's just content ourselves with showing that S_5 implies S_6. The argument will generalize without any trouble. Let's start with the inductive hypothesis. We'll fantasize that $(1 + x)^5$ is actually $x^5 + 5x^4 + 10x^3 + 10x^2 + 5x + 1$. Notice it actually is, but we're not assuming that it actually is. We're just fantasizing that it is. Now we want to look at $(1 + x)^6$, and that's going to be $(1 + x)^5$, which we already are fantasizing about, multiplied by $(1 + x)$. If you do it very carefully, you'll first get an exact copy of the first polynomial because when you multiply by $(1 + x)$, first you just multiply by 1, which gives you the exact copy, and then multiply by x, which will give you a new polynomial but with the same coefficients just shifted over with one higher exponent. If you sum the 2 polynomials, you're really just summing—when you're doing the math, you're just going to be adding the coefficients of the first polynomial and then the coefficients of the first polynomial again but shifted over. If you just look at it, you'll have a 1 to start and then a $(5 + 1)$ for your next coefficient, then a $(10 + 5)$ for the next one, then a $(10 + 10)$, then a $(5 + 10)$, then a $(1 + 5)$, and then a 1. You're doing exactly the mathematics you would do to get row 6 of Pascal's triangle. We've proven that S_5 implies S_6, and we've proven in general what's called the binomial theorem, that the nth row of Pascal's triangle is the coefficients of $(1 + x)^n$. This is an important warm-up exercise, but it's something we're going to use later.

Now we're ready, though, to get into the parity of Pascal's triangle. We're ready for a full-fledged investigation. In the real world of mathematical problem solving and mathematical circles, this is something I might do with middle school teachers or even middle school students, depending on how their attention span is. We would take maybe an hour or 2 and slow down and spend a lot of time doing pencil-and-paper conjectures. You have to remember that in this short half-hour, we're compressing a lot of mathematics, and I do urge you to try to do your investigations with that pause button because you're ready to do those investigations. We want to look at parity. What do we need to use? We certainly need to use modulo 2. We don't have to do hard calculations at all, and remember, we're starting at row 0. That's the beginning one at the top. Let's rewrite Pascal's triangle but only doing it modulo 2. The great thing about mod 2, parity, is that 1 will be any odd number and 0 will be an even number. We only need to use the numbers 0 and 1. Let's look at the first 9 rows, in other words, starting at row 0 and ending at row 8. We go 1; then 1, 1; then 1, 0, 1; and then 1, 1, 1, 1. You might think 1, 3, 3, 1, but you don't need to think about that anymore. Just look at 1s and 0s, and then the 1, 1, 1, 1 row leads to a 1, 0, 0, 0, 1 row and so on.

At first, you'll notice there are not that many evens at all. Row 4 and row 8 are all even except for the ubiquitous 1s that start and end every row of Pascal's triangle. Also, notice that row 7 and row 3, they have only 1s in them, and that allows you to make an immediate quick conjecture. You can see powers of 2 floating around in here. Row number $2^n - 1$, 1 less than a power of 2, appears to be all 1s, and the power of 2 rows appear to be all 0s, except for the first and last 1s bounding the row. Clearly, that second statement follows from the first because if you have all 1s, you can sort of think of those as little seeds that suddenly plant a row of 0s below them because $1 + 1$ is always going to give you 0. If you have a row of 1s, the very next row is going to have 1s on the end, but everything else in the middle will be 0s. How do we prove this, though? These are just conjectures. In fact, we'll do more than just prove one statement. What we'll do is look at the problem in this holistic sense. We're going to not look at single rows at a time; we're going to look at giant structures. Let's do a little more investigation. What's the proper venue for this? Do you need graph paper here? Not really. What you really need is careful pencil and paper, and one of the hard things about Pascal's triangle is that it gets big really fast. You quickly run out of room on your paper. If we want to be really correct, the proper venue of investigation here would be a computer. There's plenty of free mathematical computation software out there.

Let's look at rows 0 through 15 now. Again, you'll see a pattern where we end with 1s at the bottom, and we have these interesting triangles, inverted triangles, of 0 that are very pretty. What if we go even further to the next power of 2? If we continue and look at rows 0 through 32, we now really can see a fractal structure with inverted triangles of 0s galore. What's causing them? Again, if you have a seed of all 1s, that's going to give you a row of 0s, and then that will start a seed itself of only 0s because if you have, say, fifteen 0s, then below that will be fourteen 0s, for sure, and then thirteen 0s, and so on. If you start with a row of all 1s, you will then have an inverted 0 triangle below it. That's a pretty obvious pattern once you just look at things modulo 2. We see that the natural way to look at Pascal's triangle's parity is by these successive doublings of it, not by a row-by-row structure. What we need now is a choice of notation that proceeds not in this row-by-row way but in this kind of power of 2 structural way. Choice of notation is an important problem-solving idea. We have to think carefully about how we are talking about our problem, and that will help us to investigate it and maybe help us to prove things. Now we can begin an inductive structure. Let's use a notation T_n, standing for the nth-order triangle, one that ends with a row of all 1s. For example, T_1 is the first-order triangle, which goes up to row 1. It starts with a 1, and the bottom row is just 1, 1. T_2 would be the triangle that's twice as big. It goes 1; 1, 1; then 1, 0, 1; then 1, 1, 1, 1. And if you think about it, it is built out of 3 copies of T_1 with a 0 in the middle. That's the idea. If we keep going, we have a recurring structure. For example, if you have the T_n triangle, the one that ends with row $2^n - 1$, so it has a row of all 1s at the bottom, it will give a seed of 0s. So there will be a 0 triangle in between, but then at either end of it, you'll just have the little 1s next to a 0, and that will be a seed for a brand new T_n. You'll get a brand new copy of the top triangle. This triangle has 0s below it, and then it gives birth to more triangles. This picture encapsulates the entire inductive step because what we can do now is realize that as we double our structure, each time, we get 3 triangles of the preceding size. That's our inductive picture, $T_{(n+1)}$ is 3 copies of T_n. Now what we need to do is some careful mathematics to count the terms. We want to count the 0s, remember, because we're interested in evens. That's difficult because those inverted triangles, it's hard to measure their size. If we looked at some of the smaller examples going out just to row 15 or row 32, there are so many of these 0 triangles it's hard to keep track of them.

If counting 0s is hard, don't forget to use your important problem-solving strategy of just flipping your point of view, and let's not count 0s. Let's count 1s. If we can count 1s, that's just as good as counting 0s, and this turns out to be nearly trivial. Here's why—how many 1s are in T_1? T_1 just goes 1, 1, 1. It has exactly 3 1s. What is T_2? T_2 is 3 T_1's with a big fat 0 in the middle. T_2 is just 3 T_1's, so it has 9 1s in it. What will $T_{(n+1)}$ be in general? It will be 3 copies of T_n with a big fat 0 in the middle. If we define U_n to be the number of 1s in T_n, we have a very simple formula: $U_1 = 3$, $U_2 = 3 \times 3$, $U_3 = 3 \times 3 \times 3$, and so on; $U_n = 3^n$. We're almost there.

Our final step is to compute the relative fraction of the 1s in a T_n because, remember, we want to see what happens as this n get larger, and larger, and larger. All we need to now know is: How many members are there in T_n? How many elements are in it? That's easy because T_n is just a triangle. It has one number, then 2, then 3, then 4, so the total number of numbers in T_n is going to be a triangular number. It's going to be the number 1 + 2 + 3 all the way down to the number of elements at the bottom. How many are at the bottom of a T_n? The definition of a T_n is that its bottom row had 2^n terms. T_n is the 2^n triangular number. In other words, it's just 1 + 2 + 3 + 4, all the way up to 2^n. Don't be scared. This is an ugly-looking sum, but if you think about it, you have learned how to do this sum before. It's just the Gaussian pairing trick. We can easily write a formula for that sum. It's just $2^n(2^n + 1)$ and then the whole thing divided by 2. Now we're ready to compute a probability, at least not of the evens but of the odds. The probability that an element in the first 2^n rows is odd is equal to the number of odd ones divided by the number of terms period. That's $(3^n)/(2^n(2^n - 1/2))$, do a little fraction, bring the 2 up to the top, and we get a very scary-looking fraction. But if you stare at it, you'll see that the numerator is 2 times 3^n while the denominator is $2^n(2^n + 1)$, so it's a little bit bigger than $(2^n)(2^n)$, and $(2^n)(2^n)$ is 4^n. Despite the factor of 2 in the numerator, eventually, that 4^n will overpower the numerator. Here's a more rigorous way to see it, by dividing the numerator and denominator by 4^n. Usually, it's not a good idea to add fractions to a problem, but in a situation like this, where we want to know what happens as n gets large, we're interested in finding things that get small. If we divide the numerator and the denominator by 4^n, what we get is a fraction where the numerator is $2 \times 3^n/4^n$, which we can simplify into $(3/4)^n$, and the denominator just becomes $(1 + 2/4)^n$. As n grows larger and larger, $3/4^n$ is going to get smaller and smaller because it'll be 3/4, which is less than 1, × 3/4, which is less than 1, and so on. Likewise, $(2/4)^n$ will get smaller, and smaller, and smaller. Both of them will vanish. The denominator will approach the number 1, and the numerator will become smaller, and smaller, and smaller, approaching 0. The entire fraction approaches 0. In other words, and this is really, truly bizarre, the probability that an element is odd approaches 0. As n gets larger, and larger, and larger, that number becomes smaller, and smaller, and smaller. There are only 2 games in town, odd and even, so that means the probability that an element is even approaches 100%, which is truly surprising because if you looked at the first few rows of Pascal's triangle, most of the terms are odd, but in the long run, most of the terms are even, not just most of the terms, but closer to 100% in any number you could say. For example, if you were to go out the first 100 quadrillion terms, 99.999% of them would be even. This is a great betting opportunity if you have an infinite amount of time on your hands, and it's completely unexpected.

Even more unexpected is that you can do this with other primes and arbitrary numbers, in fact. It turns out that the probability that if you picked a number at random from Pascal's triangle, the probability that it is a multiple of, say, 2009 is going to approach 100%. Wait, there's more. Remember, we looked at the rows of Pascal's triangle, and we wanted to say that we were able to make a conjecture about how many were odd and how many were even. Can we use our knowledge of binomials to approach this problem? The answer, of course, is yes. If you looked at the table, you saw that the number of odds was always a power of 2. The question is: What's the proper venue for looking at numbers that are powers of 2? What makes powers of 2 simple? The simplest way to look at powers of 2 is if you look at them in binary, in the base 2 number system. If you remember, binary is just the number system where we use the digits 0 and 1. Instead of in base 10 where we have our 1s, and 10s, and 100s, and 1000s, and so on, now the digits stand for the 1s place, the 2s place, the 4s place, the 8s place, the 16s place, and so on, just the powers of 2. The first few numbers are 1, 1 0, 1 1, 1 0 0, 1 0 1; [for] 1 0 1, there's a 1 in the 4s place, a 0 in the 2s place and a 1 in the 1s place. The next number is 1 1 0 and so on. For example, 11 in binary is a 1 0 1 1. The first 1 is the 8s place, and then we have a 1 in the 2s place, and then a 1 in the 1s place. [The number] 15 is all 1s; it's 1 1 1 1 because it's 1 + 2 + 4 + 8. [The number] 16 is a 1 with 4 0s, a 1 in the 16s place, and so on. Binary is very easy to calculate with. The conjecture is clear. If you wrote down the row numbers in binary, you'll notice, for example, row 4 of Pascal's triangle is 1 0 0, and it has 2 odd numbers. Row 5 is 1 0 1, and it has 4 odd numbers, which is 2^2. Row 7 has 8 odd numbers, which is 2^3, and if you write 7 in binary, it has 3 1s. Our conjecture is the number of 1s in row n is equal to 2 raised to the number of 1s in n when n is written in binary. But why? Let's go back to the fact that we can look at powers with Pascal's triangle. The elements of Pascal's triangle are the coefficients of $(x + 1)^n$. For example, $(1 + x)^2$ is $1 + 2x + x^2$, but we're only interested in things mod 2, so the $2x$ disappears. In other words, $(1 + x)^2$ is equal to $1 + x^2$ (mod 2) because 2 is the same as 0 (mod 2). In mod 2, you can be a bad algebra student and still do fine.

If we square this again, we get $(1 + x)^4$ is the same as $((1 + x)^2)^2$ and that's $(1 + x^2)^2$. We'll do the same game again. We'll have a middle term that's $2x^2$, but we can eliminate it, and we just get $(1 + x)^4$ (mod 2). In other words, in general, if we take $(1 + x)$ and raise it to any power of 2, we'll just get $(1 + x)^2$. In other words, we can eliminate all the middle terms, and that immediately explains why row 2^n has just 2 1s because you only have 2 coefficients that are odd, the first and the last. What about an arbitrary one, for example, row 11? Just write 11 in binary. Remember, in binary, 11 is 1 0 1 1, because 11 is $8 + 2 + 1$. Then, let's look at row 11 of Pascal's triangle by expanding $(1 + x)^{11}$, but we won't just multiply $(1 + x)(1 + x)(1 + x)$ eleven times. We'll go $(1 + x)^8(1 + x)^2(1 + x)^1$, using this binary representation as our guide. If you look at $(1 + x)^{11}$, we know that 11, when written in binary, has just 3 1s in it. Therefore, we have 3 terms in our product using the binary representation. If we reduce this modulo 2, what we get is that $(1 + x)^{11}$ is equal to $(1 + x^8)(1 + x^2)(1 + x)$, and that's a product of 3 binomials. How do you multiply 3 binomials? When the dust settles, you'll end up with 8 different non-zero terms because if you take the first polynomial times the second polynomial, and when you do your FOIL—first, outer, inner, last—you'll have 4 terms. When you multiply that by $(1 + x)$, you will have the 1, which gives you the original polynomial, and then multiplied by x, you'll shift it over, and the total will be 8 different terms. There will be 8 different non-zero terms. Here's a slicker way to see it. When $(1 + x)^{11}$ is multiplied out and simplified modulo 2, it will be a sum of powers of x^k where the coefficients will be either 0 or 1. How do we count those non-zero terms? Just plug $x = 1$ into the polynomial, and each non-zero term will equal 1. If we plug $x = 1$ into $(1 + x^8)(1 + x^2)(1 + x)$, we get $2 \times 2 \times 2$, and of course, this will work in general. The number of 1s in the binary expansion of n will give you the same number of terms of just binomials, and when you plug a 1 in, you'll get $2 \times 2 \times 2$ that many times. That explains the phenomenon of how many odds and evens we see in a row of Pascal's triangle.

In this lecture, we used 2 important ideas to investigate Pascal's triangle. We discovered a recurrence, not of a sort of single function, like the Fibonacci series, but of a holistic, structural, fractal type. We used the clever idea of a binary representation as our natural point of view for investigating parity. These were great ideas, but the real grandeur was the scale of our investigation. We used more or less everyday problem-solving investigative tools to peer in to something that was infinite. Our next lecture, in contrast, will mostly concern itself with finite questions, but the methods will be new and unusual and powered by infinite series.

Lecture Twenty-One

Recasting Numbers as Polynomials—Weird Dice

Scope:

This is an advanced lecture that uses algebra more than most, including infinite geometric series. Can we renumber 2 dice with positive whole numbers that are not the standard 1, 2, 3, 4, 5, and 6 in such a way that the various sums still range from 2 to 12 inclusive, with the same probabilities as standard dice? Amazingly, the answer is yes. We use generating functions, which glue most of mathematics to polynomial algebra.

Outline

I. Generating functions are a method of using polynomial algebra to recast many types of problems.

A. Any sequence of numbers $a_0, a_1, a_2, \ldots$ gives rise to a generating function, the (possibly infinite) polynomial $a_0 + a_1x + a_2x^2 + a_3x^3 + \cdots$.

B. The crux idea behind generating functions is the simple observation that $x^a x^b = x^{a+b}$.

C. Here is the basic generating function strategy.

1. A problem gives rise to sequences of numbers.
2. The sequences are converted into polynomials.
3. The polynomials are manipulated in a useful way with algebra, which may tell us something about our original sequence.

II. The dice problem: Ordinary dice are numbered 1, 2, 3, 4, 5, and 6. When you roll 2 dice, the probability that you get a sum of 10 will be 3/36, since there are exactly 3 ways to get a 10 and there are 36 ways of rolling 2 dice. Our question is whether it is possible to renumber 2 dice with positive integers so that neither is an ordinary die, yet all possible sums occur with the same probability as they do with a pair of ordinary dice.

A. Suppose such dice exist. We will call them weird dice. If you roll 2 weird dice (and they may not be 2 identical dice), the probability of getting a sum of 10 will still be 3/36.

B. The denominator of 36 is not important; what matters is the number of ways to get each sum.

C. Let's label one die $a_1, a_2, \ldots, a_6$ and the other $b_1, b_2, \ldots, b_6$.

D. We want the 36 possible sums of $a_i + b_j$ to behave like ordinary dice.

E. Note that a weird die can have multiple faces with the same label—for example, three 1s, two 2s, and one 5.

F. Our problem is simple. All we need to do is look at all the possible ways to label dice and all the possible sums. A computer could do that in microseconds, but we are not computers.

G. How do we organize such masses of data? With generating functions.

III. Here are some examples of sequences transforming into generating functions or vice versa.

A. $1, 2, 3 \leftrightarrow 1 + 2x + 3x^2$.

B. $1, 1, 1, 1, \ldots \leftrightarrow 1 + x + x^2 + x^3 + x^4 + \cdots = 1/(1-x)$.

C. $1, 7, 21, 35, 35, 21, 7, 1 \leftrightarrow (1+x)^7$.

IV. Many operations are possible with generating functions, but we will stick to multiplication. Let's look at some examples.

A. Compute $(2 + x)(1 + 3x) = 2 + 6x + x + 3x^2$, using the FOIL method. There are 4 raw terms.

B. What is the coefficient of x^6 in $(x^3 + 2x^2 + x)(3x^4 + 2x^3 + x)$? We want to look at the ways we can multiply terms and get the exponent of 6. The x^3 and $2x^3$ and the $2x^2$ and $3x^4$ combine to give an answer of $1 \times 2 + 2 \times 3$, which equals 8.

V. Generating functions can shed light on combinatorics. Consider the simplest type of die (i.e., a coin). Put 0 on one side and 1 on the other. Then the generating function will be $1 + x$.

A. Suppose 7 people are each flipping a coin to decide if they will get a prize (0 = no, 1 = yes).

B. The number of prizes possible ranges from 0 to 7.

C. There are 2^7 different outcomes.

D. Each is encoded by the expansion $(1 + x)^7 = (1 + x)(1 + x)(1 + x)(1 + x)(1 + x)(1 + x)(1 + x) = 1 + 7x + 21x^2 + 35x^3 + 35x^4 + 21x^5 + 7x^6 + x^7$.

E. How many outcomes had 3 prizes? 35.

F. In other words, there are 35 ways to choose 3 prize winners out of 7 contestants. Thus $35 = \binom{7}{3}$.

VI. Now we are ready to recast the original dice problem into polynomial form.

A. The generating function for a single die is $D(x) = x + x^2 + x^3 + x^4 + x^5 + x^6$.

B. The generating function for the sums of 2 ordinary dice is just $D(x)D(x) = (x + x^2 + x^3 + x^4 + x^5 + x^6)^2$. This expands to $x^2 + 2x^3 + 3x^4 + 4x^5 + \cdots + 2x^{11} + x^{12}$.

C. If weird dice exist, we must have $A(x)B(x) = (x + x^2 + x^3 + x^4 + x^5 + x^6)^2$.

D. Now we have converted a tricky question into a relatively simple one: Can we factor $(x + x^2 + x^3 + x^4 + x^5 + x^6)^2$ in a different way?

VII. We can do this with algebraic tools.

A. Use the geometric series tool to simplify $x + x^2 + \cdots + x^6 = x(x^6 - 1)/(x - 1)$.

B. We have to factor $(x + x^2 + \ldots + x^6)^2 = x^2(x^6 - 1)^2/(x - 1)^2$.

C. We need to get rid of the denominator.

D. The full factorization thus is $x^2(1 + x)^2(1 - x + x^2)^2(1 + x + x^2)^2$, and now it is a matter of rearranging them in a nonsymmetrical way.

E. In other words, if we write each distinct prime as $P(x) = x$, $Q(x) = (1 + x)$, $R(x) = 1 - x + x^2$, and $S(x) = 1 + x + x^2$, then $D(x) = PQRS$.

F. We want to break up the product $DD = PPQQRRSS$ into 2 factors that are not the same.

1. $DD = (PQRS)(PQRS)$ does not work; it gets us back to where we started.
2. There are so many choices; how do we narrow them down?

G. The math team lemma saves the day. For each die, the sum of the coefficients must be exactly 6. So let's look at the sum of the coefficients of our prime factors.

1. $P(1) = 1$.
2. $Q(1) = 2$.
3. $R(1) = 1$.
4. $S(1) = 3$.
5. Notice that the product of these numbers is 6, as it must be!

H. So for each die, we need exactly 1 Q and 1 S, since that is the only way to get the proper coefficient sum. We can do whatever we want with the other factors, as long as each die also has 1 P.

I. We conclude that our weird dice are 1, 2, 2, 3, 3, 4 and 1, 3, 4, 5, 6, 8.

VIII. Even though we did not draw a single picture, there is a strong correspondence between what we did with generating functions and with transformational geometry.

Suggested Reading:

Gardner, *Penrose Tiles to Trapdoor Ciphers*, chap. 19.

Graham, Knuth, and Patashnik, *Concrete Mathematics*, chap. 7.

Wilf, *generatingfunctionology*.

Zeitz, *The Art and Craft of Problem Solving*, sec. 4.3.

Problems:

1. The coefficient of x^{13} when you expand $(1+x)^{17}$ is, of course, $\binom{17}{13}$. But you can also write $(1+x)^{17}$ as $(1+x)^{10}(1+x)^7$ and expand each of those factors. What binomial coefficient identity do you get?

2. Consider the infinite series $1+D(x)+\left[D(x)\right]^2+\left[D(x)\right]^3+\cdots$, where $D(x)$ is the die generating function. When this infinite series is simplified, it is the generating function for what easily stated sequence?

Lecture Twenty-One—Transcript

Recasting Numbers as Polynomials—Weird Dice

This is our second advanced lecture, and using my hiking analogy, it's going to be fairly steep throughout with a few very steep sections. Our main topic is generating functions, which is a very high-level and very powerful recasting tactic. Remember how recasting works. We recast geometry to logic with the private planets problem. We've recast numbers into geometry. We've recast numbers into combinatorial entities. With generating functions, we will recast polynomials into counting ideas and backwards. That's the idea of generating functions.

Recall the recent lecture that we had on transformational geometry. In that lecture, we got a glimpse of a wonderful and exotic problem-solving world by first developing simple lemmas about the manipulation of rotations, which we then used to solve a mysterious problem about pentagons. In this lecture, we will also expend some effort that will be steep on discovering algebraic lemmas, and we will use this to solve a mysterious problem about dice.

First let's briefly define our subject, and then we will return to the dice problem. What is a generating function? Given any sequence of numbers, possibly infinite, will form a polynomial. For example, if we start out with 1, 2, 3, 4, we'll form the polynomial $1 + 2x + 3x^2 + 4x^3$. Each sequence becomes coefficients of a polynomial starting with exponent 0, starting with a constant term. The crux idea, what makes generating functions work, is the very simple fact that when you multiply $(x^a)(x^b)$, you get x^{a+b}. In other words, there's a sort of a glue between multiplication of polynomials and addition of exponents. There's some sort of a correspondence, and we will exploit that.

The mathematician Herbert Wilf, who is a great expert on generating functions, wrote in his book *Generating Functionology,* in the very first sentence of his book, he wrote, "A generating function is a clothesline on which we hang up a sequence of numbers for display." The idea is that sequences of numbers will become polynomials and vice versa. Our overall strategy in using generating functions is: We'll take a problem, we'll turn it into a sequence, we'll turn that into polynomials, we'll manipulate these polynomials in a useful way using algebra, and that will somehow inform us about the original sequence of numbers and help us to solve our problem.

Now for the problem itself—it's the problem of the weird dice. Ordinary dice are numbered from 1 to 6, as you know, and when you role 2 dice, you add up the 2 numbers. For example, that's a 1 and a 3. We get a 4, and in general, the probability that we'll get a particular outcome depends on how many ways we can get that outcome. There's only one way that we could get a 12, for example, rolling 2 6s, and there's only one way we could get a 2, rolling 2 1s. But if we wanted to get a 7, it turns out there are 6 ways you could do that. You could get a 1 and a 6. You could get a 2 and a 5, and you could get a 3 and a 4, a 4 and a 3, a 5 and a 2, or a 6 and a 1. There are 6 different ways you could get a 7, and since there are 36 ways that we could roll these dice, that would be 6 out of 36.

The denominator doesn't really matter, so let's just keep track of the number of ways it can happen. Here's our question. Our question is: Is it possible to renumber the dice with positive integers? Imagine these are the blank slate. Could we renumber them with positive integers so that neither of them is an ordinary die. By ordinary die, I don't mean the silly thing about how the 1 and the 6 have to be opposite. I mean the numbers are completely different. They're not the numbers 1 through 6. That's what a weird die is. I take 2 of them; they are possibly different, but could I have 2 weird die so that neither is an ordinary die, yet I could mathematically argue my way into a casino and say these are perfectly legitimate to use to play craps because the probabilities will be the same? In other words, the probability of getting a sum of 2 would be 1 because there's only one way they could give me a 2. There's only one way they could give me a 12, but there'll be 6 ways they could give me a 7. That's my problem. Can we do it?

Let's go back and look at ordinary dice. We know that the sum of 2 can happen in only one way. The sum of 3 can happen in only 2 ways, and then it goes up to 3, 4, 5, 6 ways for 7. Then it goes back down symmetrically. We have this nice little chart—1, 2, 3, 4, 5, 6, 5, 4, 3, 2, 1 for all the different sums. That's exactly the chart we want to get with our weird dice. Suppose we labeled the dice with 6 variables. Say we call one die the A die and the other one the B die, and we'd have our variables, a_1, a_2 through a_6 for the A die and b_1 through b_6 for the B die. There are 36 possible sums. There's $a_1 + b_1$, $a_1 + b_2$, $a_5 + b_3$, all these different possible sums, but we want those sums to have exactly the same distribution as ordinary die. We want there to be only one sum that adds up to 2, only 2 sums that add up to 3, and so on.

A weird die, as long as it's not an ordinary die, could have multiple faces with the same label. For example, there's nothing wrong [with] having a weird die where my die maybe has 3 1s on it, and maybe you could even have a number like 9. It doesn't have to have the numbers from 1 to 6. Here's an example: Suppose our weird dice were a die with 1, 4 2s, and a 3, so it's a 1, 2, 2, 2, 2, 3 die. The other one, the *B* die, is 1, 2, 3, 3, 5, 9. It has 2 3s, but it goes all the way up to 9. Let's verify some sums here. How do you roll a 2? The only way you could roll a 2 would be if the *A* die landed with a 1 and the *B* die landed with a 1. There's only one way it could happen. How about a sum of 3? Verify that there are 5 ways that could happen. The *A* could give you a 1, and the *B* could give you a 2. There's only one way that could happen, or any of the 4 faces of the *A* that's a 2 could be paired with the face of the *B* that's a 1. That's 4 different ways, so there's a total of 5. A sum of 5 would happen in 8 different ways because you could have a 3 from *A* and a 2 from *B* just in one way, but you could have any of the 4 2s in *A* matched with any of the 2 3s in *B*. It's one of those simple menu problems. You have 4 choices of picking the 2 face in *A* and 2 choices of picking the 3 face in *B*. There will be 4 × 2, or 8, different possible ways you could end up getting that sum. The total would be 8 ways of a 2 and a 3 plus one of a 3 and a 2, a total of 9 different ways of getting a 5. That's how weird dice are.

Our problem is simple. All we need to do is look at all the possible ways to label dice and look at all the possible sums and see if we could get one that's not ordinary dice. A computer could do that in a few microseconds, but we are not a computer. Our question is: How do we organize such masses of data? The answer is generating functions. That's the purpose of this lecture, but how, and why, and what? Let's see some simple examples of generating functions in action. The idea of a generating function is to turn a sequence into a polynomial. For example, the sequence 1, 2, 3 becomes the polynomial $1(x^0) + 2(x^1) + 3(x^2)$. The infinite sequence 1, 1, 1, 1, 1, 1 … becomes the infinite series $1 + x + x^2 + x^3 + x^4$ and so on, but we know the formula for that infinite series. It's just $1/(1 - x)$. That's already showing you that you can take an infinite sequence and replace it with just a finite term, very useful possibly. You could take the Fibonacci numbers and write them as a generating function,

$x^1 + x^2 + 2x^3 + 3x^4 + 5x^5$ and so on. The coefficient of x^n is the nth Fibonacci number. More relevant to us, we could look at Pascal's triangle, the sequence 1, 7, 21, 35, 35, 21, 7, 1. We can rewrite that in generating function language as the polynomial $(1 + x)^7$. We've seen some generating function ideas in action in the previous lecture.

As with our geometric transformations lecture, there are many algebraic things you can do with generating functions, but we'll stick to just one, just to multiplication; just like with transformations, we've mostly just looked at the composition of rotations. Let's look at polynomial multiplication again. You've gotten a little bit of practice in the last lecture, but let's review. Remember how FOIL works when you multiply 2 binomials. For example, $(2 + x)(1 + 3x) = 2$, the first terms; $+ 6x$, the outer terms; $+ 1x$, the inner terms; $+ 3x^2$, the last terms. Those are the raw terms, and you can then simplify them if you wanted to and add the 2 middle terms. In general, we're interested in the raw terms at first. For example, how many raw terms are there in just a product $(1 + x)(7 + x)$? Forget about simplifying. You have 2 terms in the first, 2 terms in the second, so there will be 4 different raw terms before you simplify.

Here's a harder question: What's the coefficient of x^6 when we multiply the 2 polynomials $(x^3 + 2x^2 + x)(3x^4 + 2x^3 + x)$? Well, let's think about it. How do we get an x^6? We get it by adding exponents. Look at the exponents and find the ones that add up to 6. We have an x^3 in the first. We can pair it with the $2x^3$ in the second one, and that will give us a coefficient of 3. We also have an x^2, which we can pair with the x^4, and that gives us a coefficient, when you do the multiplication, of 6, so I believe we'll get $6 + 2$ so far, and there are no other ways we could get an x^6. The idea is we look for the things that sum to a 6 and find out all those combinations and add them up.

Now we'll ask the same question: What's the coefficient of x^6 for this uglier product? If you look at that uglier product, $(x^3 + x^2 + x^2 + 1)(x^4 + x^4 + x^4 + x^3 + x^3 + x)$, you realize that it's actually the same as before but just written in terms of raw terms. The answer would be exactly the same, but it reminds you that you can break down things into raw terms and look at them in this expanded-out form or in a more compact form.

Here's something that we used in the last lecture, which I want to give a name to. I'll call it the math team trick. Here's a typical sort of junior high school math team question. Find the sum of the coefficients of the polynomial $(2+3x-4x^2)^3$. The hard way would be to expand out that polynomial, begin with an 8 and end with $-64x^6$, and then you could just plug 1 in in order to add those coefficients. Instead, just plug 1 into the original thing. Just plug 1 into $2+3x-4x^2$, and what do you get? You just get 1; $1^3=1$. The answer is 1. The math team trick is plug 1 in and you'll get the sum of the coefficients. That allows us to prove in a flash some binomial identities that we discovered back, I think, in Lecture Four when we were just beginning our discovery of Pascal's triangle. Remember the sum of the rows was 2^n for row n. That's easy. We know that the elements in row n are the coefficients of $(1+x)^n$. Just add up those coefficients by plugging 1 in, and we get $(1+1)^n$, which gives you 2^n. Remember the alternating sum, that if we just alternate adding and subtracting, we got 0. Well, that just comes from plugging in $x=-1$ into the expansion of $(1+x)^n$; $(1-1)^n$ is just 0, so instantly we're able to prove these complicated conjectures.

Let me reiterate the lemmas that we've discovered. The first one I call the math team lemma, which says that the sum of the coefficients of a polynomial is found by just plugging 1 into the variable. The other one is the multiplication lemma, which tells you how to find the exponent in a product. Just look for all the terms where the exponents sum to the one you want, and you multiply all of those individually and count them up. We'll look at that in a little more detail because we are interested, remember, in algebra itself. We're interested in applying our algebra to the problem at hand.

Let's reformulate this back to dice. Let's imagine abstract dice, and here's a very simple case where when I say abstract dice, they don't even have 6 spaces. Let's keep it simpler. Suppose die A has the numbers 1, 3, 3, and 4, and die B has the numbers 0, 1, 2, and 3. Then we're going to turn them into generating functions, so I'll have the function $A(x)$; I'll let it equal $1x^1+2x^3+x^4$. The coefficient of, say, x^3 is the number of 3s. The coefficient of x^4 is the number of 4s. There is no coefficient for x^2 because there are no 2s in it. Likewise, $B(x)$ is the polynomial $1x^0$, or just $1, +x+x^2+x^3$. If you read that off, you can see that B has one 0, one 1, one 2, and one 3. Now what happens if you wanted to roll these 2 dice? Suppose you want to roll die A and die B, and you want to find out how many 4s you will get. Well, the way you'll get it, if we just go back to the original, die A could have a 1 corresponding with a 3 from die B, or you could have the 2 3s from die A with the 1 from die B. Instead, the way to do it is to take a look at the polynomial $A(x)$ and the polynomial $B(x)$ and just multiply them. If you multiplied $A(x) \times B(x)$ and you just look for the coefficient of x^4, it will automatically be counting the number of ways you can take exponents from A and exponents from B and get them to add up to 4. The total number will be exactly the number of combinations you'll get. When you multiply the coefficients, it's like doing that menu problem. You'll get the exact answer you want. I call this the multiplication lemma.

Let's look at the simplest possible example, which is coins, and I understand it takes a while to wrap your brain around this concept. Imagine a 2-sided die, which we'll call a coin. Put a 0 on one side and a 1 on the other. The generating function, then, is just $1 + x$. Suppose you had 7 people—Adam, Betty, Carlos, Dave, Erika, etcetera, and you have them flip the coins to decide if they will get a prize. If you get a 0, it's no, and if you get a 1, it's yes. The number of possible prizes will range from 0 to 7, and the number of possible outcomes will be 2^7 because there are 2 outcomes for each person. You just multiply them, $2 \times 2 \times 2$, 7 times. Each outcome, though, is encoded in the product $(1 + x)(1 + x)(1 + x)$, 7 times, $(1 + x)^7$. If you wrote that out, you get $1 + 7x + 21x^2 + 35x^3 + 35x^4 + 25x^5$ and so on. That immediately allows you to find out all kinds of things. How many outcomes will have 3 prizes? That's the x^3 coefficient, 35. In other words there are 35 ways to choose 3 prizewinners out of the 7 contestants. In other words, 35 is that binomial coefficient $\binom{7}{3}$.

Let's go back to our original problem. We're ready to reformulate it into this polynomial form. Here's the generating function for a single die. It's really rather simple. A single die will have one 1, which is an x, and it will have one 2, which is an x^2, and likewise, an x^3, x^4, x^5, and an x^6. All the coefficients are 1, and the other die will have exactly the same generating function. Let's look at the product. That will give you the distributions of the sums. We take $x + x^2 + x^3 + x^4 + x^5 + x^6$ and we square it. Well, indeed, when you do the algebra, you get $x^2 + 2x^3 + 3x^4 + 4x^5$ and so on. The coefficients exactly match. For example, the coefficient of x^7 will be 6. If weird dice exist, they will have their own generating functions. There'll be the A die with $A(x)$ and the B die with $B(x)$, but what we know is when we multiply those 2 polynomials, we'll get exactly the same symmetrical 12th-degree polynomial that we get when we multiply 2 ordinary dice. What we've done is we have now turned the problem into an algebra problem. All we need to do is take a look at that giant 12th-degree polynomial and figure out if we can factor it into a funny way that's different from the ordinary. But what remains now are just algebraic tools. What we need to do is factor that 12th-degree polynomial.

Let's first just look at one die, $x + x^2$ all the way to x^6, and we can use the geometric series formula to write that as $(x^7 - x)/(x - 1)$, and we can factor out an x from the numerator. We have $x(x^6 - 1)/(x - 1)$. We have to multiply that by itself and factor it some more, and we have to get rid of the $x - 1$ in the denominator. We take $x^6 - 1$, and we can write that as a difference of 2 squares, since x^6 is $(x^3)(x^3)$, so we can write $x^6 - 1$ as $(x^3 - 1)(x^3 + 1)$. If you haven't studied algebra in a while, this might seem pretty hard and it is hard work, but I want to stress that at this point, this is technical work. This is stuff that can be done with a high school algebra textbook. It's completely standard.

Then we factor these cubics; $x^3 - 1$ can factor into $(x - 1)(x^2 + x + 1)$, and $x^3 + 1$ factors into $(x + 1)(x^2 - x + 1)$, and what's great is now we can factor out those $x - 1$'s in the denominator and we have complete factorization where we have duplicated terms. We have x^2, we have $(1 + x)^2$, we have the polynomial $(1 - x + x^2)^2$, and then finally $(1 + x + x^2)^2$. Now it's a matter of rearranging them in a non-symmetrical way. Let's be organized. Let's write each of these distinct sort of prime parts. We'll call them P, Q, R, and S. For example, $P(x)$ is the polynomial x. $Q(x)$ is $(1 + x)$. $R(x)$ is $1 - x + x^2$. And $S(x)$ is $1 + x + x^2$. So the dice polynomial, the ordinary dice, $D(x)$, is $PQRS$, and what we want to do now is break up the product $DDPPQQRRSS$ into 2 factors that aren't the same, because the original factorization $(PQRS)(PQRS)$ won't work because it gets us back to where we started. All we have to do is take those letters $PPQQRRSS$ and break them up. There are so many choices. How do we do it? Don't forget that math team lemma.

What do we know has to be true about weird dice? No matter how weird they are, they have only 6 faces, which means the sum of the coefficients has to be 6. For example, if you had a weird die that had 3 1s and 3 5s, it would be the polynomial $3x + 3x^5$. The coefficients would still add up to 6. We know that the coefficients add up to 6, and the coefficients must all be positive. Let's focus on the sum of the coefficients of these little prime bits, P, Q, R, and S. It's easy to check. $P(1) = 1$ because it's just the polynomial x. $Q(1) = 2$ because it's $x + 1$; plug a 1 in. $R(x) = 1$ because it's $x^2 - x + 1$. We still just get a 1, but $S(1) = 3$ when you plug a 1 in. Notice that the product of those numbers is 6, as it has to be, because that's a single die.

So for each die, we need exactly one Q and one S because that's the only way to get the proper coefficient sum, no matter what, whether they're ordinary die or weird die. We can do whatever we want with the other factors. Or can we? Each die still needs one P. Remember, P is the polynomial x. We need that because otherwise, we would end up with polynomials beginning with the number 1, which is $1x^0$ and we don't want any 0 labels. Both dies must have PQS, and we're done because if both dies have to have PQS, then the only possibilities are $PQRS$ and $PQRS$, which gives us the original dice, or PQS versus $PQRRS$.

Let's examine that factorization and see what it gives us. If we just look at PQS, we have $x(1 + x)(1 + x + x^2)$, and that gives us $x + 2x^2 + 2x^3 + x^4$. Then if we look at $PQRRS$, we have $x(1 + x)(1 + x + x^2)(1 - x + x^2)^2$. When you do all the algebra in that—and by the way there are tricks to do that easily, but we could just let a computer do that for us if we want—we get $x + x^3 + x^4 + x^5 + x^6 + x^8$. Now we can read off the coefficients and conclude that our weird dice are 1, 2, 2, 3, 3, 4 and 1, 3, 4, 5, 6, 8. For example, if you look at that first polynomial, the coefficient of x^1 is 1. The coefficient of x^2 is 2, which means there are 2 2s. The coefficient of x^3 is 2, which means there are 2 3s, and so on. So the weird dice we read off from the coefficients are 1, 2, 2, 3, 3, 4 and 1, 3, 4, 6, 5, 8.

Here's one die that has an 8 in it, and here is the die that has 2 2s and 2 3s. Let's verify that they work. Well, we'd have to really go to a casino for a few hours, but let's just do a few rolls. Here I got a 7, and it turns out there will be 6 ways of getting that 7. How about snake eyes? There's only one way to get that. How about getting 12? With ordinary dice there's only one way to do it. The way you do it here is you have to be lucky enough to roll a 4 with one and an 8 with the other, $4 + 8 = 12$, and there's only one way that can happen. These weird dice will solve the problem, and this is an amazing example of recasting dice into numbers, numbers into polynomials, doing algebra with polynomials, getting us back into numbers, back to dice.

Let's step back a little and think about what we did. In this lecture, we didn't draw a single picture, but there's a very strong correspondence between what we did with generating functions and what we did earlier with transformational geometry. Both were tactics informed by strategy, flipping a point of view, in the case of geometry, and recasting, in the case of this lecture. Both used algebraic lemmas that were pretty hard and maybe a little dry, but in both cases, once we were comfortable with our lemmas, we could apply them to investigate and solve really unusual, offbeat problems. With both of these ideas, there's a richness hinted because we're just scratching the surface of these tactics. It's clear that you can use them in a very wide variety of problems. For example, if you let generating functions become infinite, you can use the ideas of infinite series.

There's one more thing I want to talk about. The actual problem we solved and the solution we got is really interesting and significant because this is perhaps the first case in our course where we got a genuine asymmetrical solution. For example, in the "Squarer Is Better" lecture, where we looked at optimization, the answer always turned out to be: to optimize, symmetrize. And that's what intuition tells you to do. This is a case where we had 2 possibilities, the symmetrical answer, yet hidden, there was this asymmetrical answer. In one sense, we have an interesting discovery, namely, that symmetry isn't always the answer because this is clearly not symmetrical compared to this. These are the same, and these are wildly different, different from each other and different from the original dice. At the same time, if you looked at these dice from a very abstract point of view, this dice was *PQRS*, both of these were *PQRS*s, and here, what we had was a *PQS* and a *PQSRR*, where the 2 *R*s were brought over. In a certain higher-level way, the 2 *R*s put together, even though it was a binomial squared, $(x^2 - x + 1)^2$, because the sum of the coefficients was 1, you could think of that as almost like a coefficient whose size and absolute value is 1. You could say, these are completely symmetrical, sort of like 2 electrons, and these, even though they don't look symmetrical from a higher-order way of just thinking about sums of coefficients, it's as though this one is an electron and this is a positron. There's still a little bit of symmetry left. It's good to think about these higher-level ideas because, after all, as you’ve seen, symmetry is such an important idea in all of problem solving. Even when things aren’t symmetrical they have symmetry.

Lecture Twenty-Two

A Relentless Tactic Solves a Very Hard Problem

Scope:

This advanced lecture is a continuation of the ideas begun in Lecture Fourteen. We use the pigeonhole principle relentlessly to study Gallai's theorem, a Ramsey-style assertion. Our investigation takes us into the realm of the nearly infinite, where we contemplate numbers far larger than the number of atoms in the universe. The strategic principle we highlight is more earthbound: Don't give up.

Outline

I. Our focus will be on finding structure within seemingly randomly colored lattices. Our basic tool is just the pigeonhole principle and the notion of coloring.

- **A.** If you 2-color the points on a number line, you only need to look at 3 points, and you are guaranteed that 2 are the same color.
- **B.** If you have a 3×3 grid, there are 2^9 ways to color it.
- **C.** If you were using 10 colors, it would be 10^9.

II. Warm-up problem: Color the lattice points of the plane in 2 colors. Prove that there must be a rectangle (with sides parallel to the axes) each of whose vertices are the same color.

- **A.** The pigeonhole principle applied to 3 consecutive lattice points in a horizontal line forces there to be at least 2 points of the same color (points are pigeons, colors are holes).
- **B.** We would be done if we had 2 identical patterns, one on top of the other. But how do we guarantee this can happen?
- **C.** The crux idea: Look at 9 rows of 3 points, and we are guaranteed that 2 of the rows will be identical!
- **D.** After all, there are only 8 different 2-colorings of 3 points!
- **E.** And since each row contains 2 points of the same color (at least), we will have a monochrome rectangle.

F. Thus every 3×9 grid of points contains a monochrome rectangle. This is a worst-case scenario, of course; we could have gotten lucky with just a 2×2 grid, but 3×9 guarantees it.

G. This can be easily generalized. If we 5-colored the lattice points of the plane, there would still be a monochromatic rectangle, but we would need to look at rows of 6 points (to guarantee that each row contains at least 2 points of the same color) and then consider $5^6 + 1$ (or 15,626) rows.

III. An innocent, obvious generalization is can we get a monochrome square? This is called Gallai's theorem: If you 2-color the lattice points, you are guaranteed to find a monochrome square. We will proceed as we did with the rectangle and build up our square in stages.

A. First we need a monochrome line, then a monochrome isosceles right triangle, and finally a monochrome square.

B. The line is easy: Just pick any 3 points, and at least 2 of them must be the same color.

C. But how do we get the rest of our right angle? How can we control the distance between the points of the same color?

IV. We need at most 3 points to get our 2 monochrome points, so our 2 monochrome points lie in at most a 3×1 grid.

A. If we are to build a right angle with this as a starting place, we will need a 3×3 grid.

B. With 2 colors and 9 points in the grid, there are 2^9, which equals 512, possible 3×3 grids.

C. Thus, if we looked at a row of 513 3×3 grids, at least 2 of them would be colored in exactly the same way.

D. And we are guaranteed that each of these grids will have 2 points of the same color at the top. We could have a monochrome right angle in a 3×3 grid, but we may not. The worst-case scenario is that among our 513 grids, 2 are the same, but neither have right angles.

E. Wishful thinking says to build a right angle! We do not have control over the colors in the third 3×3 grid, but we can focus on the point on the lower left.

F. This was a worst-case scenario. We are guaranteed that in any 1539 × 1539 grid, we must have a monochromatic right angle.

G. Let $R(c)$ equal the size of the grid that guarantees a monochrome right angle if we use c colors. So $R(2) = 1539$. Define $S(c)$ to be the size of a c-colored grid that guarantees a monochrome square. We wish to prove that $S(2)$ is finite.

V. How do we create a monochrome square? Certainly the monochrome right angle is a start. What if we could make a monochrome right angle out of monochrome right angles?

A. The worst-case scenario is that the lower right corner is not red.

B. In this case, we can guarantee a structure where 3 of the 4 vertices are red but the fourth vertex is blue. Then we can again make a right angle with these structures.

C. Then we are done; no matter what color the lower right-hand point is, we have created a monochrome square.

D. However, in order to do this, we needed a right-angle construction with identical right angles. We can only get monochrome right-angles so far using 2 colors.

E. But what if we could get right angles using any number of colors? That is what we need to complete our proof.

VI. We know that any 2-colored 1539 × 1539 grid is guaranteed to have a monochrome right angle.

A. There are only $B = 2^{(1539 \times 1539)}$ different ways to 2-color this 1539 × 1539 grid. This is a number with 712,996 digits. It is unimaginably larger than any "real" number.

B. There are about 10^{80} particles in the universe. If every one of those particles could count at a speed of a billion billion billion billion billion numbers per second, it would take the universe 10^{712870} seconds, or 10^{712863} years, just to count this number.

C. Suppose it were possible to get a monochrome right angle if we color the lattice in B colors. In other words, there is some size $G = R(B)$ such that any $G \times G$ grid that is B-colored is guaranteed to have a monochrome right angle.

D. Assign a different color to each of the B possible ways to 2-color a 1539×1539 grid.

E. Now view the entire 2-colored lattice, but in 1539×1539 chunks. You can think of this as a B-colored lattice. If $R(B)$ is the finite value G, then there is a grid of $G \times G$ chunks that is guaranteed to contain a monochrome right angle of chunks!

VII. Thus, if we can show that $R(B)$ is finite, we are done. It was already pretty hard to compute $R(2)$, but we will confidently compute $R(3)$, and do it in a way that can clearly be generalized to higher numbers of colors.

A. Our guiding principles are to stick to worst-case scenarios and build structures with focal points.

B. We start by considering a row of 4 points. This guarantees 2 identical colors somewhere in this row.

C. There are 3^{16} different ways to 3-color this 4×4 grid, so just string $3^{16} + 1$ of them together in a row, and we are guaranteed to see 2 identically colored grids. Let $M = 3^{16} + 1$.

D. Next, focus on a grid that would contain the third vertex of our right angle. Since it is a worst-case scenario, it is possible that this focal point is colored green.

E. Thus, we have shown that in any $4M \times 4M$ grid of 3-colored points, we may not have a monochrome right angle, but we at least are guaranteed a structure like the one we built: an almost, hopeful monochrome right angle.

F. How many different $4M \times 4M$ grids can there be? Just 3^{16M^2}. This is a very large number, which dwarfs the superbig number B. Let's add 1 to this number and call the sum G, for gigantic.

G. If we place G of these $4M \times 4M$ grids in a row, we are guaranteed to get 2 that have the same color. But remember that every $4M \times 4M$ grid is guaranteed to have an almost monochrome right angle structure.

H. Once again, we focus on the hopeful lower left corner. Whether it is green, blue, or red, we have a monochrome right angle.

I. Clearly, it is possible to keep doing this, creating mind-boggling structures until we manage to guarantee a monochrome right angle for B colors.

Suggested Reading:

Graham, Rothschild, and Spencer, *Ramsey Theory.*

Soifer, *The Mathematical Coloring Book.*

Problems:

Recall that in Lecture Fourteen we used the pigeonhole principle to prove that if we color the plane in 2 colors, no matter how we color the plane we are guaranteed to have 2 points that are the same color and are exactly 1 meter apart. Here are 2 problems that prove stronger things with this same hypothesis. Use the worst-case scenario methods from this lecture.

1. Color the plane in 2 colors. Prove that 1 of these colors contains pairs of points at every mutual distance. In other words, 1 of the 2 colors, say, red, has the property that for each distance x, there are 2 red points exactly x units apart. (Hint: Use proof by contradiction.) What is the negation of the assertion?
2. Color the plane in 2 colors. Prove that there will always exist an equilateral triangle with all its vertices of the same color.

Lecture Twenty-Two—Transcript

A Relentless Tactic Solves a Very Hard Problem

What can we accomplish if we use the pigeonhole principle absolutely relentlessly, and I really mean relentlessly? In this lecture, our focus will be to look for structure within seemingly randomly colored lattices. Our goals will be modest. We will look for rectangles, each of whose vertices are the same color, and then later squares, monochrome squares, where each vertex is the same color. Our goals are modest, but our methods are not modest. Our methods are outrageous. Our methods will be bold. This is the third advanced lecture, and to use the hiking metaphor, it's very steep. It's steep throughout, and it goes along a cliff edge most of the time. What I would encourage you to do is to view this lecture, but you might have trouble with it at first because along the way, we'll be using mind-bogglingly large structures and even larger numbers. It'll be a little hard to understand at first. If you want to skip this, it's not essential for the remaining lectures. If you want to skim this and come back to it later, that's okay, but I hope you find it interesting. Again, the theme is: What can we do if we just try to use a single tactic and just never give up? When I say never give up, I really mean that.

Let's start with some basic tools because we've already studied these coloring problems, these simpler Ramsey-style problems. Let's review the basic idea. Suppose you are 2-coloring the points on the number line, in other words, just coloring them red and blue, say. Then, you only need to look at 3 points, and you're guaranteed that 2 are the same color. That's the simplest version of the pigeonhole principle. On the other hand, if you had a grid, say, a 3×3 grid, of lattice points, there are 9 points in it. Then there would be 2^9, $2 \times 2 \times 2$, 9 ways, which is 512, 2^9 different ways of coloring that one little grid. If you had 10 colors, it would be 10^9. The numbers get pretty large pretty quickly, and what we'll be doing is redefining the notion of color. We'll sometimes look at the different ways of coloring a grid, and we'll think of grids as colors in their own right, as you'll see.

Our warm-up problem for this advanced lecture, to give you an idea of the conceptual difficulties, is a problem from the USA Mathematical Olympiad. It's a simple coloring problem, at least it seems simple, to look for a rectangle. Here's the problem that I've adapted: Color the lattice points of the plane in 2 colors. Prove that there must be a rectangle, with sides parallel to the axes, each of whose vertices are the same color. This is our first step.

Using the basic principles, what we'd like to see are some repeated patters, just 2 red dots and then underneath them 2 red dots. If we just look at 3 consecutive lattice points in a horizontal line, we're guaranteed to have at least 2 points of the same color. Let's call them red without loss of generality. There will be a sequence of maybe red-blue-red, or red-red-blue, or blue-red-red, or maybe red-red-red, something like that, but we're certainly guaranteed to have 2 of the same color in any 3-point sequence. We'd be done if we had 2 identical ones right underneath each other, but we can't guarantee that can happen. If you have 3 dots here colored in some funny way, underneath them anything could happen. The crux idea is to look now at 9 rows. Why 9 rows? Because if you have 3 dots, how many ways can they be colored? $2 \times 2 \times 2$, 2^3, which is 8. There are only 8 different 2-colorings of the 3 points. If you looked at 9 of these rows, by the pigeonhole principle applied again, you're guaranteed to have 2 rows that are the same. For example, in this picture, here are 2 rows that are the same, and we're guaranteed, remember, that if you have any row of 3 dots, 2 of the dots in that row are the same color. So we have 2 rows that are the same, but we're guaranteed that each row has 2 of the same color. If you look at those 2 identical rows, they actually form a red rectangle. This is the basic model of what we're going to do. We can extend this idea some more; let's do so.

We've proven that in any 3×9 grid of points where we're coloring with 2 colors, we're guaranteed to have a monochrome rectangle, where all the 4 vertices are the same color. It might happen all the way at the boundaries of that 3×9, or it might happen right up at the corner. It doesn't matter; we're guaranteed somewhere in that grid of 3×9 to have our rectangle. That's a worst-case scenario. We could have had it happen a little bit sooner, and it's possible we could have always done it with a smaller grid, but that doesn't matter. This method is easy to generalize, and we can generalize it to more colors.

Suppose we 5-colored the lattice points. Could we still find a monochromatic rectangle? Sure, we could. We'll use exactly the same idea, only now what we would start out with is a row of 6 points. That would guarantee 2 points the same color. Let's use that as sort of a miniature lemma. Every row of 6 points will have 2 the same color, and so now all we need to do is figure out how many rows we need so that we're guaranteed 2 of those 6-point rows are going to have themselves the same coloration. How many ways can you color 6 points using 5 colors? $5 \times 5 \times 5 \times 5 \times 5 \times 5$, which is 15,625. If you had 15,626 rows, then 2 of those rows would have to have rows of 6 points that have exactly the same coloration. For example, you might have 2 rows colored blue-red-green-yellow-black-red, another one blue-red-green-yellow-black red, and we're guaranteed to have a rectangle, maybe a very, very, very skinny rectangle, but all of whose vertices are red. We can do this with any number of colors. That's our background.

Here's the thing I would like to accomplish in this lecture—ramp up the idea from rectangles to squares. Here's our modest increase. I'm just going to ask very politely: Suppose we 2-colored the lattice points on the plane, so red and blue, is it possible to guarantee a monochromatic square, where all 4 vertices are the same color? This is called Gallai's theorem, which was proven in the middle of the last century. It seems like it can't be much harder than a rectangle. A square is more symmetrical than a rectangle. It should be nicer. The question is: Is it possible no matter how the lattice is colored to get a monochromatic square? You might want to just try to mess around and look at this rectangle method to see what you can come up with. You'll find that it's surprisingly hard to do, but it can be done. In order to do it, we're going to need awesomely gigantic numbers and heart-stoppingly relentless use of wishful thinking and the pigeonhole principle. What we're going to do is proceed in the same way we did with the rectangle, and we're going to build things up in stages. With the rectangle, we started with a row, and then we found another row. With a square, we'll start out with a row, 2 points that are the same, a monochromatic line, if you want to call it that, and then in the next stage, we're going to build a right triangle, an isosceles right triangle, in other words, a triangle of 3 points, 1, 2, 3, where the legs are the same size, so it's 3 points of a square. And then, the next stage, if we can accomplish that—it's sort of our intermediate step—is to then try to build a square.

Let's start with the easiest part, the line. That's easy. All we need are 3 points, and we've got 2 of them of the same color by pigeonhole. No problem there, but how do we get the rest of a right angle? How can we control the distance between the points of the same color? For example, with 3 points we could have red-blue-red, but then we would need to have a red point down here, exactly the same distance away as this red and this red. How do we guarantee it? Let's not worry exactly. The idea is we'll start to build patterns, and we'll build patterns inside patterns. Our strategy is to build a pattern that's almost right, look at worst-case scenarios, put it inside an even larger structure, and then create out of smaller, almost correct structures a single large, correct monochromatic structure.

We'll start with those 2 monochrome points, say, 2 red points, and we needed at most 3 points to do this. We have them in a 3×1 grid. If we're going to build a right angle with that as our starting point, we'll need a 3×3 grid. Now, how many ways can you color a 3×3 grid? Remember, we worked that out already. There are 2^9 possible colorations of those 3×3 grids. Imagine now—$2^9 = 512$—so imagine, 513 3×3 grids—grid, grid, grid, grid, grid, 513 of them. You're guaranteed that 2 of them will look exactly the same. They'll have exactly the same color scheme using the 2 colors, and we're also guaranteed because they're 3×3, that each of these grids will have 2 points of the same color at their tops. We could have been lucky. It could have been that the 3×3 grid in the first place had its right angle in it. That would have been great, but we have to deal with the worst-case scenario. The worst-case scenario at the moment is among our 513 3×3 grids, 2 of them are the same, but neither of them have monochrome right angles. The worst-case scenario looks like this. You have your 2 identical 3×3 grids, and they're separated. This diagram is not drawn to scale. They could be as far as 511×3 units apart, but what we know is that these are 3×3 grids. It's a worst-case scenario, so that third point isn't red. The 2 top ones are red, but the bottom of the angle has to be blue. This is our worst-case scenario. The worst-case scenario is we got this.

What do we do now? We want to build a right angle, so wishful thinking says, build a right angle. Just imagine this configuration where the top could be as far apart as 513×3, and the bottom is the same distance. Inside this grid, we'll have the top, we'll have 2 things that have this almost right angle, and then we just look at the third one, which we color in gray because we don't know what the colors are supposed to be. Just think about those colors. We don't have any control on them, but it doesn't really matter. Look at the lower left corner. I call that the focal point of this structure. It's either red or blue. If it's red, then you'll notice that you get a right angle with point A on the top and point B all the way at the right of our structure. We have a perfect right angle if it were red, but it might not be red. It could be blue, but if it's blue, then we get a right angle with the lower left of the 2 structures on the top. We found a right angle. The way we did it was by having these almost right angles coordinate in such a way that when you look at this focal point, this mystery point, it has a lot of possible choices, but no matter what color we give it, it ends up creating the structure we want, namely, a right angle, 3 points at a right angle that are all the same color.

This was a worst-case scenario, but what we're guaranteed is that in any 1539×1539 grid, that's 513×3, a giant, giant grid, we're guaranteed that there will be 3 points that form this monochromatic right angle. We've accomplished a monochromatic right angle, and if we wanted to, we could say, "Let's pretend that's Gallai's theorem and now we can go home." But it's not. We need that fourth point. Let's use some notation. Let's call $R(c)$ to be the size of the grid that guarantees a monochrome right angle if you use c colors. What we've proven so far is that $R(2) = 1539$. If we use the letter S for squares, what we want to do is compute $S(2)$ and prove that it's finite. $S(2)$ would be the size of a grid in order to guarantee a monochromatic square, 4 vertices that are all the same color. So far we've got a right angle, and that's pretty good.

Let's think about a penultimate step. How do you create a monochrome square? Certainly, a monochrome right angle is a start. What if you could make a monochrome right angle out of monochrome right angles, in other words, a configuration of 3 monochrome right angles that are all the same? That's kind of like a monochrome right angle of monochrome right angles. Here's a picture of it. We have 3 red right angles, and they themselves form a right angle. We have these 3 grids that are all the same, and you'll notice that these are not 3×3 grids. They could be much larger. Remember, in order to guarantee right angles, you need to have 1539×1539, but to have a structure like what we're looking at, who knows how big it is? Let's just use wishful thinking. Let's suppose we got that structure. There it is. The natural wishful thinking thing to do is to look at that fourth corner. Let's take a look. The identical red right angles don't form squares, so in a worst-case scenario, the 3 little right angles—if we were lucky, they'd form a square right then and there—but if they didn't, then they would have to have blues in their lower right corners. Let's assume they do, and now we have a beautiful focal point. If you look at point A, either it's blue, in which case we have a blue monochrome square using the 3 blues in the corners of our red right angles, or that point A is red, in which case if you look at the outer squares of our configurations, we have 4 red squares that are all evenly placed. Again, if we could get this configuration of a monochrome right angle of monochrome right angles, then we'd be done.

Well, how do we do that? We're able to get a monochrome right angle when we 2-colored the plane, but how do we get a monochrome right angle of monochrome right angles? How do we do something like that? The idea is: What if we could get right angles, not using 2 colors but getting any number of colors? Let's again use wishful thinking. Suppose no matter how we colored the lattice points, we're guaranteed to get a monochrome right angle. If we colored with a million colors, we're guaranteed that somewhere there will be like a fuchsia, fuchsia, fuchsia or something else because we have a lot of colors to choose from. We know that any 1539×1539 grid with our ordinary 2 colors is guaranteed to have a monochrome right angle. How many of these 1539×1539 grids are there if we're just 2-coloring them? That's easy. It's just $2^{(1539 \times 1539)}$. In other words, $2^{2,368,521}$, which is itself a number with 712,996 digits. This number, the number of ways you could color the 1539×1539 grids just with 2 colors, is unimaginably larger than any number that has anything to do with the real world. For example, there are only 10^{80} particles in the universe. If every one of those particles could count at a speed of a billion, billion, billion, billion, billion numbers per second, it would take the entire universe collectively $10^{712,870}$ seconds just to count this number, just to count it. But it's still a finite number, and the point is that's the total amount of diversity in 1539×1539 grids. This number is really, really big. Let's call it B for big. Suppose it were possible to get a monochrome right angle if we colored the lattice in B colors. In other words, there's some size. We'll call it G equaling R(B) such that in any $G \times G$ grid that is B-colored, we're guaranteed to have a monochrome right angle. Remember, B colors transcends physical reality. There are not enough colors in the spectrum. There are not enough particles in the universe, but it's still a mathematical idea.

Assume that if we B-color the plane, we'll get these 3 points making a right angle. Now, assign a different color to each of the B possible ways to 2-color a 1539×1539 grid. In other words, view the entire

2-colored lattice, what we've been handed with, but in 1539×1539 chunks. Each of these chunks, think of them as like a single pixel now that we can color in any one of the B colors. We can think of that as a B-colored lattice, and if $R(B)$ indeed is a finite value, some number G, then in this giant $G \times G$ grid of chunks of 1539×1539 pieces, we're guaranteed to have a monochrome right angle, where we'd have a 1539×1539 chunk, another chunk, and another chunk that themselves have the same coloration. In other words, in the 2-colored situation, they are identical. This is all wishful thinking. If we can show that $R(B)$ is finite, then we'd be done because we would be able to find a right angle, right angle, right angle that's exactly the same. It was already hard enough to compute $R(2)$. How do we get $R(B)$ where B is that unimaginably large number? We'll be modest. We're thinking inductively here. We're going to do the inductive step. We're going to go from $R(2)$ to $R(3)$. We're going to find the size of a 3-colored grid that guarantees a monochrome right angle. The way we'll do it will be easy to generalize. If we can go from $R(2)$ to $R(3)$, you'll see that we could go to $R(4)$, $R(5)$, and to any number we want. We'll be done if you can understand how to construct a monochrome right angle using 3 colors. Our guiding principles are: Let's stick to those worst-case scenarios, but let's build these recursive structures that hold where we start with an almost solution and then we make a larger structure of almost solutions that forces an actual solution.

Let's try it. Take a deep breath, and remember, we're hiking along a cliff today and we don't look down. We take a deep breath, and let's start with our 3-coloring. Consider a row of 4 points, 1, 2, 3, 4. This is the easy part. By the pigeonhole principle, we're guaranteed that there are 2 identical colors somewhere in this row. We have our line. Now we wishfully think about the potential right angle this makes. Remember, it's a worst-case scenario, so the worst possible thing we might see is just 2 red points on the top, inside a 4×4 grid here, with unfortunately, a different color at the bottom. We'll call that different color blue. We won't give up. We're going to try to repeat what we did earlier. If you look at this, this is now a 4×4 grid. It has 16 possible lattice points in it, but there are 3 colors. There's 3^{16} different ways to color it, to 3-color this 4×4 grid. Imagine just stringing 3^{16} plus one of these 4×4 grids together, and we're guaranteed to see 2 identically colored grids. Let's call this number, $3^{16} + 1$, M. M is a tiny number. It's only 43 million, approximately.

Again, here's a picture that's not to scale, and remember, it's a worst-case scenario. We know that, without loss of generality, each of those grids contain 2 red points at the top, equally spaced, but then making the right angle is not a red point. It's a different colored point, but it's the same different colored point. We have red-red-blue squares that are maybe millions of units apart. We don't have a right angle yet, but we know what to try. We know where to look next. Where to look next is that third corner. Look at that square, and we'll try to do the same game we did before. Let's use that focal point. Remember, with 2 colors, if we just looked at that lower left point, we'd be done because either it's red or blue, and no matter how we do it, we'd end up getting ourselves a right angle. But we have 3 colors, so in the worst-case scenario, we didn't have a red or a blue point down there; we had a green point down there. Because we have 3 colors to choose, the worst-case scenario is that we have red-red-blue on top, another red-red-blue on top, and then down in the lower left corner we have a green. Well, it's not what we wanted. That's a worst-case scenario, but it is, at least, a structure.

An ordinary problem solver at this point would give up, but we're not going to give up. We know that we're on a steep, steep trail, and we're almost there. All we need to do is just do it again. Remember that number M is about 43 million. In any $4M \times 4M$ grid of 3-colored points, we may not have a monochrome right angle, but we are at least guaranteed that sort of funny structure, that hopefully almost monochrome right angle, the red-red-blue, red-red-blue, green structure. That's the worst-case scenario.

Let's play the game again. How many $4M \times 4M$ grids could there be? Only 3 raised to the $(4M \times 4M)$, 3 raised to the $16M^2$. This is a very large number. It has 82 million digits. It dwarfs that super big number, B, which had less than a million digits, but it won't hurt to add 1 to this number. Let's take 3 to the $16M^2$ and add 1 to it, and let's call it G, G for gigantic. G is 1 more than the number of possible $4M \times 4M$ grids. Remember, this is what a $4M \times 4M$ grid contains. If we place G of these $4M \times 4M$ grids in a row, we're guaranteed to get 2 of them that look the same because they have the same coloration. Remember, every $4M \times 4M$ grid in the universe is guaranteed to have one of those almost monochromatic right angle structures. This is what we're going to see: We're going to see a red-red-blue, red-red-blue, green structure and then very, very far away, maybe $4MG$ units away, another one that looks exactly the same.

You could think of these as just super-superstructures. What do we do now? We make it into a square, of course. At the top—it's possibly $4MG \times 4MG$, an unimaginably large-sized grid—the top, remember, we have 2 identical almost right angle structures that contain the colors red on top, and the little blues on the lower left of the top squares, and then the lower left are 2 greens. Then in the bottom, we have nothing we know about yet, but let's focus, as always, on that lower left point. It's the focal point because it's the point that makes a right angle in more than one way. That's the key. The symmetry of this picture is designed so that we can create perfectly symmetrical right angles with different dimensions.

Focus on that lower left-hand corner. It could either be red, green, or blue. If it's green, then it matches with the 2 lower-left greens up at the top to form a green right angle. If it is blue, it matches with the lower-left blue on the top and the lower-left rightmost blue on the top to create a blue-blue-blue right angle. Finally, if it's red, we get a red-red-red right angle using the outermost points in our structure. With a $4MG \times 4MG$ grid, we were guaranteed to have a monochrome right angle using 3 colors. In other words, we did it. We proved that $R(3)$, the size of the grid that guarantees a monochrome right angle, is equal to $4MG$ where, remember, M is about 43 million, 4 is 4, and G is a number with, I think, 82 million digits, but it's a finite number, unimaginably larger than any number you'd ever, ever imagine. Even if the universe had B particles in it counting, it couldn't count to G, but clearly, it's possible to keep doing this by just creating mind-boggling structures. Instead of superstructures, we'd create super-duper-structures. Then we'd create super-duper-superstructures and so on and so forth. It's easy to turn the 3-color situation into a 4-color situation and so on. If we wanted to be formal, we'd have to write an induction proof, but we don't need to because we've already drawn the picture. The conclusion is we can create our right angle in any number of colors. Remember, that was the penultimate step for creating a 2-colored monochromatic square. We don't know the size of the grid, but we know that it's going to be finite. We succeeded in proving one version at least of Gallai's theorem, which actually can be generalized to arbitrary shapes.

This was about as close as we can get to infinity. We solved a really hard problem by calmly and bravely repeating ourselves, looking for structures and superstructures and using again, and again, and again the same wishful thinking/pigeonhole argument on these larger and larger structures. The larger lesson is a metaphorical one. There is no limit to what we can achieve if we harness the infinitely compressible universe of mathematics and couple it with the irrepressible imagination of a good problem solver, especially a brave problem solver. In the next lecture, we will encounter an infinite structure, yes, but the crucial aspect will not be infinite structures or infinite processes but just a single crux move, a very, very ingenious crux move. That will lead us to also examine briefly the nature of creative genius and the adulation of this genius.

Lecture Twenty-Three

Genius and Conway's Infinite Checkers Problem

Scope:

In our penultimate lecture, we sketch John Conway's brilliant solution to a classic puzzle. Our focus is not just on the mathematics, which is a wonderful mix of the ubiquitous golden ratio and monovariants, but we also engage in a discussion of mathematical culture, particularly the cult of genius that surrounds Conway and other mathematical "rock stars," including Paul Erdös and Évariste Galois.

Outline

I. The checkers problem: Place checkers at every lattice point of the half plane of nonpositive y coordinates. The only legal moves are horizontal and vertical jumps. By this, we mean that a checker can leap over a neighbor, ending 2 units up, down, right, or left of its original position, provided the destination point is unoccupied. After the jump is complete, the checker that was jumped over is removed from the board. Is it possible to make a finite number of legal moves and get a checker to reach the line $y = 5$?

- **A.** It is easy to get to $y = 2$, and with work, we can get to $y = 3$. It is reasonable to conjecture that we cannot get to $y = 5$.
- **B.** What methods do we have for proving impossibility?
  1. Come up with a quantity that can be calculated for each configuration.
  2. This quantity should be a monovariant.
  3. If the quantity, say, always decreases but needs to increase in order to get to $y = 5$, we would be done.

II. Conway's monovariant: Using the coordinate system with all checkers at $y = 0$ and below, let the target point be $C = (0, 5)$. We wish to prove that we can never reach this point.

- **A.** Define the number $z = (-1+\sqrt{5})/2$.

B. For each point in the plane, compute its "taxicab distance" d to the target point C. For example, the point (2, 1) has distance 2 + 4 = 6.

1. Then compute the value z^d.
2. For each configuration of checkers on the plane, add up the values z^d for each point that has a checker. This will be an infinite series. This is the Conway sum for that configuration.

C. For example, the entire first row ($y = 0$) has the Conway sum $z^5 + 2(z^6 + z^7 + \cdots)$.

1. This simplifies to $z^5 + \dfrac{2z^6}{1-z}$.
2. Since $z^2 + z = 1$, we simplify this further to $z^5 + \dfrac{2z^6}{z^2} =$

$$z^5 + 2z^4 = z^3(z^2 + z) + z^4 = z^3 + z^4 = z^2(z + z^2) = z^2.$$

D. Likewise, the Conway sum for the row $y = -1$ will be z^3, and so on, so the starting Conway sum of our problem is $z^2 + z^3 + z^4 + \cdots = \dfrac{z^2}{1-z} = \dfrac{z^2}{z^2} = 1$.

III. Why is this a monovariant? Consider any configuration of checkers, and look at what happens to the Conway sum when a jump occurs.

A. For example, suppose there are checkers at (4, 1) and (4, 2), and (4, 3) is unoccupied, so the first checker can jump over the second.

1. Before the jump, there is a checker at a distance of 8 and one at a distance of 7.
2. Afterward, the checker with a distance of 8 is now at distance 6, and the 7 checker is gone.

B. So the Conway sum changes by the net amount of $z^6 - z^7 - z^8 = z^6(1 - z - z^2) = 0$. In other words, if a checker jumps toward the target point C, the Conway sum does not change!

C. Consider a jump away from the target. Suppose a checker is at distance 10 and jumps over a checker at distance 11 to end up at distance 12. Then the net change is $z^{12} - z^{11} - z^{10} = z^{10}(z^2 - z - 1) = z^{10}(1 - z - z - 1) = z^{10}(-2z)$, which is negative.

D. Thus if you jump away from C, the Conway sum decreases.

E. There is one other case to check: that where your jump does not change the distance to C. For example, if you jump from $(-1, 2)$ to $(1, 2)$.

   1. In this case, we start with 2 checkers, one at distance d (the jumper) and one at distance $d - 1$ (the jumpee). After the jump, the jumpee is gone, and the jumper is still at distance d.

   2. So the net change is $z^d - (z^d + z^{(d-1)}) = -z^{(d-1)}$, which is again negative.

F. Thus, the Conway sum is a true monovariant, never increasing from its initial value of 1.

G. All that remains is to note that if we ever were to get a checker to C, the Conway sum would be larger than 1, since $z^0 = 1$ would be supplied by C, and there would still be infinitely many other checkers to add up.

H. But our starting value is 1, and the Conway sum is a monovariant. So we can never reach C!

IV. It takes a certain type of intellect to solve problems at this level. The key ingredient is a passion to investigate without any worry about consequence.

V. Conway is one of a triumvirate of such heroes that also includes Paul Erdös and Évariste Galois. All 3 are iconoclastic rebels, belying the myth of the boring nerd, who supply romantic inspiration for the next generation.

VI. John Conway has led an unconventional life and made incredible contributions to math. He is like an eternal child in his ability to play, break rules, work on whatever pleases him, and continually ask questions, with a willingness to get his hands dirty.

VII. Paul Erdös led a life of deliberate homelessness and celibacy. He wrote more papers and collaborated with more people than any mathematician in history.

VIII. Évariste Galois died in a duel at age 20. His 60 pages of mathematics are considered by some to be the most important 60 pages ever written in mathematics. His greatest achievement, now called Galois theory, is a point of view flip.

IX. All 3 of these people had passion, commitment, and a willingness to investigate relentlessly. This is something that we can all strive for, even if we cannot all possess genius.

Suggested Reading:

Berlekamp, Conway, and Guy, "The Solitaire Army," in *Winning Ways for Your Mathematical Plays.*

Hardy, *A Mathematician's Apology.*

Hoffman, *The Man Who Loved Only Numbers.*

Honsberger, *Mathematical Gems II*, chap. 3.

Problems:

1. The solution to the checkers problem was pretty subtle. Test your understanding: Why not just assign a large value—say, 100—to the point *C*? Then if a checker occupied *C*, the Conway sum would be at least equal to 100. But since the Conway sum starts at the value of 1 and never increases, it can never reach a value this large and hence never occupy *C*. What is wrong with this argument?
2. Here is a puzzle about Erdös numbers. (Assume, for simplicity, that when mathematicians write joint papers there are only 2 collaborators.) There are 5 mathematicians in a room. Each of them has written a paper with at least 1 of the others in the room. Exactly 1 of them has written papers with 3 of the others in the room, and exactly 1 has written papers with 2 others. One of the 5 mathematicians is Erdös himself. What are the possible Erdös numbers of these 5 people?

Lecture Twenty-Three—Transcript

Genius and Conway's Infinite Checkers Problem

This is the penultimate lecture of our course on problem solving. By now, you may not be an expert, but you've seen many, many ideas and explored many problems, some quite difficult and elaborate. But no course on problem solving is complete without a look at Conway's checker problem. It's a fantastic example of creative, fearless analysis of a game, and it's a mainstay of mathematical circles and competitive problem-solving teams.

No discussion of problem-solving culture is complete without a look at Conway himself, whom you've heard bits and pieces about in earlier lectures. Conway is one of a few mathematical rock stars that are worshiped by nerds worldwide. I will conclude this lecture with a brief look at culture again, and this time, what I'll focus on is the nature of mathematical genius from a problem-solving perspective. We'll look at Conway, some lesser lights, and some perhaps greater lights, as well. First, let's take a look at the mathematics.

Here's the checker problem: Imagine that you have an infinite checkerboard, and you're placing checkers at every square of each row on a half plane. Imagine that this goes on forever all the way in this direction, checkers going this way, and this way, and this way. We're placing them on lattice points, so think of this as the x-axis. This is $y = 0$, and this is $y = 1, 2, 3, 4$. The checkers are on all the negative points. The only legal rules are horizontal or vertical jumps. In other words, I can jump this checker over this one, and then I remove that checker. Then I could jump this one here and remove it. I could do another jump, as well, and so on, a very simple game. It's a game not played with 2 players but a game that you play by yourself. It's a solitaire game, and the object of the game is to get to level 5. This is $y = 1, 2, 3, 4, 5$. For example, we can go like that and then go there. Then I could go maybe like that and like that. Then, boom, boom, boom, and I got all the way up to level 2. Remember, I have lots of checkers over here, and I could probably get something over here, gather my forces, and go boom, boom, boom. I got all the way up to, I think, level 3, but I want to get to level 5. The question is: Can you do it and, if so, how? You've seen many problems in this course where the question is can you do it and, if so, how. This is known as Conway's checker problem. You saw we could get to level 2 and we could get to level 3, but can we get to level 5? If you worked on it for a long time, you would have a lot of difficulty, and in fact, you wouldn't be able to do it. If you spent a while playing this game, you might unconfidently guess that it's impossible. That indeed proves to be the case, but what methods do we have for proving impossibility if we think of a penultimate step? This is a hard problem, but it's a tactical problem.

The solution here is to use a monovariant. What we want to do is come up with a quantity that can be calculated for each configuration of checkers. Remember, there are infinitely many checkers. We like this quantity to be a monovariant. If the quantity, say, always decreased, but it needed to increase in order to get to $y = 5$, then we'd be done. That's the sort of thing we'd look for, but what should our monovariant be?

This is a lecture where I'm not so much interested in doing an investigation with you as much as presenting a solution to you because the thing that we come up with is not something that the average person would invent on their own. This is Conway's idea. Let's use the coordinate system that we started. We have our levels $y = 0$, $y = 1$; we have the negative y's. But we also should put in an x-axis, and it doesn't matter where it goes, so without loss of generality, let's just say there's a point with an x coordinate of 0 and a y coordinate of 5. We'll call that point c. We'd like to try to get to that point c, and we're going to prove that we can never reach it. Our target is the point (0, 5).

Let's consider a number. How about the number z defined to be -1 + (the square root of 5)/2? This number is a little bit strange, but it's not the strangest number in the world because it's the reciprocal of the golden ratio. It's approximately 0.618. What are some properties of this number z? Well, first of all, it's positive, and it's less than 1 because it's 0.618, approximately. It also obeys a quadratic equation. In fact, that's where it came from. That was its inspiration. It obeys the quadratic equation $z^2 = 1 - z$. In other words, $z^2 + z - 1 = 0$. If we applied the quadratic formula to that, we would get the value $z = -1 \pm$ the square root of 5/2, and we're just picking the positive root. It's this weird number.

Now it's time to define our monovariant. For each of the infinitely many points in the plane, what we'll do is compute its so-called taxicab distance d to the target point c. What I mean by taxicab distance is just the distance along the lattice point where you're not allowed to go diagonally, as if you're a taxi driving on the city streets. For example, from (2, 1) to our target point c at (0, 5), we have to go north 4 blocks and over 2 blocks. The taxicab distance would be 6. Taxicab distance is easier to compute than ordinary distance. It's always an integer. Just add the x and y displacements. For each point on the plane, there is a distance to the target by taxicab distance, and then for each of these points, compute the value z, that mystery number that we defined, raised to the d power. For each configuration of checkers on the plane, just add up those values of z^d for each point that has a checker. Notice this is going to be not just an infinite series but a doubly infinite series because there are going to be infinitely many checkers on rows, and there will be infinitely many rows if we go down into the $-y$ values.

For example, if we start out where the checkers are in the initial position, so they're all on that first row $y = 0$, there are infinitely many checkers. What would the value be there? What we want to do is compute z^d for each value. Well, the one that's right underneath c, the origin itself, is going to be a distance of 5 from it. It's going to be z^5, and then the ones on either side are z^6, z^7, z^8, z^9, and so our sum is going to be $z^5 + 2(z^6) + 2(z^7) + 2(z^8) + 2(z^9)$ and so on, an infinite series. How do we compute this infinite series? It's easier than you think because we know the formulas for infinite geometric series. These are just powers of z, and z, remember, is less than 1, so the infinite series converges. That sum, just of that single row, is $z^5 + 2$, and we use our formula $a/1 - r$, so it's equal to $2(z^6)/1 - z$.

Now we'll use the fact that $z^2 + z = 1$, and we can simplify this further into (z^6/z^2), which is $z^5 + 2z^4$, and we can factor out a z^3 from that. Eventually, if we use the relationship that $z^2 + z = 1$, we end up getting that the entire row will be z^2 in size. This infinite series simplifies just to z^2. That's row 0. What about the row below it? That's easy. Everything there has a taxicab distance just one more than the starting row. So if you have a point over here in the starting row, the point right below it, it's taxicab distance to the target is just one more. Likewise, this point here, its taxicab distance is one more than this one. Every point in the line $y = -1$ is going to have a taxicab distance one more than the person above it. If we added all of them up, it's as if we just took that first row and just multiplied it by z to increase the exponents by 1. The value for that row, $y = -1$, will just be z^3, and then the next one will be z^4, and then z^5, and so on. If we added up the entire half plane, the entire initial configuration, it would be $z^2 + z^3 + z^4$ and so on. That's a new geometric series whose first term is z^2. So it's $z^2/1 - z$, but that's the same as z^2/z^2 because $1 - z$ is z^2, and then we get 1. In other words, this weird number z was cleverly designed so that when we added up this doubly infinite series, going out in both directions and then going down to negative infinity, when we added up all of these infinitely many exponents, we got the value of 1. The starting value of this thing is 1, and we'll call this entire thing, where we add up all of the z^d's for all the checkers in our configuration, we'll call that the Conway sum. So the Conway sum starts out at 1. Now, why is it a monovariant? So far, we've just computed it for the starting value, and at least it has a nice starting value.

Consider any configuration that occurs later. What we want to figure out is how it evolves. What happens to the Conway sum when a single jump occurs? For example, suppose there's a checker at (4, 1) and at (4, 2), and (4, 3) is unoccupied, so you're able to do the jump, and the first checker can jump over the second checker. Before the jump, there was a checker at a distance of 8 and then another one at a distance of 7. Afterwards, the checker with the distance of 8 is now at distance 6 because it's moved toward the target. The distance 7 checker was jumped, so it disappears. The Conway sum changed by the net amount of z^6, sort of the new value, but we lost z^7 and we lost z^8, so the net change, if we think about new minus old, is $z^6 - z^7 - z^8$. Well, factor z^6 out of that, and we're left with $1 - z - z^2$, z^6 times that quadratic polynomial. But we know that $1 - z - z^2$ is 0. In other words, if a checker jumps toward the target point—there's only 3 checkers that are involved here—when we compute the net change in the Conway sum, the net change is 0. Maybe the Conway sum is an invariant. That would be an interesting thing, but it's not quite an invariant.

Consider a jump that goes away from the target. For example, suppose a checker is at distance 10, and it jumps over a checker that's at distance 11. If the target's up here and I'm over here, I jump away like that, and I end up at distance 12. Then the net change is going to be z^{12}, the new amount, $-z^{11} - z^{10}$, the ones that disappeared. When we factor z^{10} out of that, what we're left with is $z^{10}(z^2 - z - 1)$, and that gives us, if we simplify, $z^{10}(1 - z - z - 1)$ and we end up with $z^{10}(-2z)$, which is negative. So it's either standing still or it is decreasing. Thus, if you jump toward c, the Conway sum doesn't change, and if you jump away, it will decrease. There's only one other case to check, and that would be where your jump doesn't change the distance to c. For example, if c is over here, maybe you're doing a jump like this, where you're on one side of the y-axis and you move to another side of the y-axis. For example, if you jumped from $(-1, 2)$ to $(1, 2)$. In this case, we start with 2 checkers. One is at distance d, and the other one is at, say, distance $d - 1$—that's the jumpee, and d is the jumper. After the jump, the jumpee is gone, but the jumper is still at distance d. Well, the net change has to be negative because the jumper still has the same value, but the jumpee disappeared, so we just lost some power of z. Again, the net change was negative. We could compute it explicitly, but there's no need. It's just going to be negative. Therefore, what we can conclude is that the Conway sum is a true monovariant. It never increases. It either stands still or it decreases, and its initial value is 1.

All that remains is to note that if we ever were to get a checker to the target point c, what would the Conway sum be? At the target point c, what's the distance to c? The taxicab distance is 0. What is z^0? It's 1. If any checker ever occupied the target point c, you'd have at least a checker with a distance 0 contributing 1 to the Conway sum. Then there'd be a whole bunch of other checkers, infinitely many, because we have to do this in finitely many steps, so there are going to be infinitely many checkers left over. They're going to add something to the Conway sum. If we got to the point c, the Conway sum would be bigger than 1, but the Conway sum starts at 1 and it's a monovariant, so it can never get bigger than 1. Problem solved—we can never reach c.

This is the sort of problem that when you see the solution to it, you go, "Wow! Incredible," but you also probably say how did Conway come up with this out of thin air? He must have been a genius. Well, Conway is a genius, but let's not canonize him yet because the solution of this problem can be broken down into several stages. First of all, we recognize that this is a typical arena in which monovariants are productive. It's a game, a finite game, and we're trying to prove impossibility. That's one of the typical sorts of situations where you look for invariants or monovariants. We look for some measure of the game and see as the game evolves if there is a value that's increasing or decreasing. We look at starting points and ending points. Once you start hunting for a monovariant, you've won half the battle. The clever idea was to say: Could this be a monovarient question? Next, you happily settle for a quantity that sometimes is invariant as long as when it does change it only changes in one way. That's what a monovariant is, and since we need to evaluate infinite configurations, one thing to try is a number less than 1 that will converge when we use an infinite geometric series. It's kind of natural to think of exponents. I'm not saying the problem is easy or that Conway wasn't creative, but I'm trying to break it down into steps to show you how an experienced problem solver might approach this.

Finally, the fact that a jump towards c takes 2 checkers with a distance d and $d-1$ and leaves you with a single checker at a distance of $d-2$, it leads you to think about sort of a Fibonacci-style situation. where you have 2 things that differ by 1. Then the third one is at the next stage. Like you have an a, b versus $a+b$ or something of that sort or a 1, 2, 3 distance situation. I'm being a little vague here, but an experienced person with, say, generating functions, when they see something like this, they tend to think about a generating function, where you might have x^2 and $x+1$ interacting with one another, or maybe x^2 and x interacting with 1, or something of that sort. It naturally lets you think about equations of that sort, and again, if you're experienced, you say, "Oh, maybe the golden ratio is involved because that is $x^2 = x + 1$." It's not quite the golden ratio, but once we start thinking along those lines, we're getting very close to the right value of z.

This is an incredibly hard problem involving problem solving, and you can think of this as one of the pinnacles of how even a competitive problem solver learns about mathematics. At some point in the stage of someone's training, they tend to learn about the Conway checker problem, and they tend to think, "Wow, Conway is amazing." They learn about other cultural icons, as well. But let me change the perspective a little bit. Instead of talking about the highest-level people of the Conway stature, let's look somewhere in between Earth and heaven and look at, for example, the teenagers who I helped to train back in 1994 who were on the International Math Olympiad team, the IMO team, from the United States. We called this the "Dream Team" because that year the United States won the Olympiad with a perfect score, and no team has ever done that, ever before or ever since. They were extraordinarily successful mathletes. They were the highest-order mathletes of the land at that time.

These kids learned about the Conway checker problem when they were fairly young, and they worshiped Conway. My question is: How different are they from you, for example? I'm not trying to insult anybody here, but let's assume that you haven't won as many math contests as some of these teenagers did. What's the difference between them and an ordinary person who loves math? There's not as much difference as you think. When I was helping to run the training program, what I noticed about these people is they were certainly very smart, but the biggest difference between them and an ordinary high school student was their ability to work. In other words, it's kind of a boring answer, which you find true in many endeavors. If you look at the greatest classical pianists, they are usually known for how much they practice. If you look at Tiger Woods, he's certainly somebody who practices golfing a lot. The people who start early with lots and lots of practice tend to be quite good at this, and they tend to like doing it.

In our Olympiad training program, the regimen would work like this: The students would have 4 hours of lecture in the morning, during which they'd usually sleep, but if I called on them, they would give a good answer. They would have about 4 hours of class, and then, almost every other afternoon, they would take a 3-hour test to simulate the International Math Olympiad, where they'd work for 3 hours on maybe just 3 problems, very, very difficult problems. In the evenings, they would have a fair amount of free time, but they'd also have a lot of homework. What would they do during the evenings? Well, they would mostly do math because it was fun. They'd do math together. They'd get into arguments about problems. They'd work on homework, and they would play bridge and they'd play Frisbee. The important thing is these people were doing mathematics for 10 to 12 hours every day, and then every night, they would break curfew. They wouldn't break curfew to go out drinking. They would usually break curfew to just hang around and play pinball or read math books. This was at Annapolis Naval Academy where the training program was, and I'd get up in the middle of the night sometimes to go to the bathroom, and I'd see 2 or 3 of the team members just sprawled out on a bare floor, sleeping with a math book over their heads. It looked very uncomfortable, but it made me happy to realize that they fell asleep doing mathematics. The idea here is what makes you great, at least on one level, is loving mathematics and wanting to work on it as much as possible.

Now let's go up to a higher level. Let's look at some of these icons, and in particular, I just want to briefly talk about what I think of as a triumvirate. Conway is one of them, and the other is Paul Erdös, who died in 1996. The last one is Evariste Galois, who died in 1832. Only Conway is alive of these, and he's approximately 70 years old. All 3 of them have a lot in common. First of all, they truly are worshiped not by a religion but in a sort of informal way. When you come up in your training, you just learn about people like this. You tend to unconsciously or sometimes even consciously emulate them. All 3 are iconoclastic rebels in some sense. They are nerds, but they're not boring nerds. They supply romantic inspiration in many ways. Also, all 3 are men, and they're all white men, but that's a whole other story. It's one, frankly, that's changing very, very quickly. Ten years from now, I might talk about 3 different cultural icons.

Let's start with Conway a little bit. Whenever I introduce Conway when he comes to the Bay area to give a talk, I love to introduce him and say that in my opinion, this is the greatest person to ever come out of the city of Liverpool, England, John Lennon being the second greatest. Conway is very pleased to hear that, and I'm sure he agrees with me. Conway is truly a remarkably imaginative person, and he does not have any false modesty. He considers himself to be a very smart, imaginative guy. He has led an unconventional life. He has been married a million times, and he does whatever he wants. He's not the sort of person who you can just send an e-mail to who will respond to you. He truly lives, within the math world, a sort of a rock star life. He does what he wants, and if people want to get in touch with him, they'll go to Princeton, where he works, and find him. He has made incredible contributions to just about every area of mathematics: to group theory, to knot theory. He has invented a new theory of symmetry. He has made fundamental contributions to quantum mechanics. He has discovered a new system of numbers involved with viewing games. But he is best known to the lay people and even to contest math people for his recreational math contributions. He was the inventor of the Game of Life, one of the first cellular automata. He was an important player with the Rubik's cube back when it came out. And if you remember that look and say problem, 1, 1, 1, 2, 1, and so on, only Conway would have looked at that problem and realized there was real, genuine mathematics involved in it. He was able to come up with a theorem about the Look and Say sequence that involved a 73-degree polynomial equation, probably the highest-degree equation ever found in a recreational math article in history.

Conway is an eternal child. What makes him a great mathematician is that he loves to play, and he loves to play with what he wants to play with. He gives back to the problem-solving community. He visits math camps every summer and spends lots of time with young people, and he's always nice to them. He shows them literally how important it is to get hands dirty. He's very interested in the Fibonacci series, and he knows that pinecones tend to have Fibonacci numbers in them. His hobby is collecting pinecones that do not have Fibonacci numbers in them. He has determined empirically that this happens in one out of a thousand pinecones; he spends a lot of time literally getting his hands dirty, just doing dirty experimentation just for fun. He epitomizes the problem-solving qualities of concentration, confidence, and creativity, but it's his ability to play and it's his ability to come up with things out of thin air that the students love to emulate.

Let's turn now to 2 others. I just want to briefly talk about the other members of the triumvirate. If you think of Conway as sort of the court jester, then there's the more serious one, and that's Erdös, who lived over 80 years but he led a completely unconventional life. Unlike Conway, who has been married several times, Erdös was deliberately celibate all his life. He was also deliberately homeless. He never held a conventional job. He traveled around the world. He was a mathematical monk with an unwavering and complete commitment to mathematics. His specialty was problem solving itself. He worked on hard problems, often problems of the Gallai theorem variety. In fact, Gallai was a good friend of his. He focused on finding elementary methods that you could apply to complex problems. He used to call something "The Book," where you would have the most perfect, most elementary, most elegant proof of every theorem. He used to say that you don't have to believe in God, but you have to believe in The Book. Erdös, who never married, was the most sociable of mathematicians. He collaborated with over 500 people, and there's something called an Erdös number. Erdös is given the number 0, and if you wrote a paper with Erdös, you get a number of 1. If you wrote a paper with someone with an Erdös number of 1, you get a number of 2, and so on. There are probably a million people with finite Erdös numbers. Hank Aaron, the baseball player, has an Erdös number of 1, in fact, because he and Erdös both signed a baseball when they were both getting an honorary degree, I think, in Atlanta.

Along with Erdös, who has this sort of monk-like persona, where all he does is math but he loves every minute of it, there is sort of the anti-Erdös, and that was Galois. Galois was French. He was handsome, and he was romantic. The most romantic thing about him was that he died at age 20, in a duel no less. He had a complete antipathy toward authority, but like Conway, he had a divinely inspired creativity. In the last year of his life, he put together about 60 pages of mathematics that some mathematicians consider perhaps the most important 60 pages ever written in mathematics. His specialty was asking the right questions, and what he did was a fantastic point of view flip. He looked at polynomials, looked at the roots of polynomials, and rather than looking at the roots, he looked at the transformations that permute those roots. He looked at abstract permutations, developed something called group theory, and created a way of analyzing polynomials to determine when you could solve for them with a finite number of algebraic steps, like the quadratic formula. This branch of math called Galois theory is still at the research frontiers today, and this was done by someone who died at age 20.

The math team kids, whether at the IMO level or at a middle school, they know about these 3 people, and they revere them. I don't want you to think that this is some kind of a cult, and I don't want you to think that they're intimidated by this genius because the most important thing that they revere, and this is something that I hope you will revere, too, is you don't want to be like Conway, you don't want to be like Erdös, you don't want to be like Galois, except for one thing. You do want to emulate, if you can, what made them special, and what made them special is passion, commitment, and also willingness to play. If you can do those things with your mathematics, you might not reach the same achievements of, say, an Erdös or a Galois, but you'll reach some achievements that you'll find very, very satisfying, and you'll be in touch with the beauty of mathematics itself. That's all you can do: Play hard.

Lecture Twenty-Four
How versus Why—The Final Frontier

Scope:

In this final lecture, we look back on what we have learned, talk about what we should study next, and reflect on what we do not know. We begin to ponder the ultimate purpose of an investigation: the quest for *why* something is true, not just how. I will share some of my favorite examples of this elusive intellectual quest.

Outline

I. First, some reminders about how to approach problems tactically, with the assumption that by now you have internalized key strategies.

- **A.** Proof by contradiction should be used when the thing you are trying to prove is easier to contemplate when negated.
- **B.** The extreme principle is useful when your problem has entities that become simpler at the boundary.
- **C.** The pigeonhole principle works well when the penultimate step can be formulated with 2 things belonging to the same category.
- **D.** Use induction when your problem involves recursion.
- **E.** The most important thing is to ask what type of problem you are trying to solve. This is half the battle.

II. What should you study next? Complex numbers!

- **A.** They are another playground with incredible potential for connecting many branches of math.
- **B.** Complex numbers are 4 things simultaneously: numbers, locations in the plane, vectors, and transformations!
- **C.** Complex numbers allow you to recast to and from physics, with dynamic interpretations of hitherto static objects.
- **D.** Complex numbers have connections to every branch of math, including number theory.

III. The birds-eye view of problem solving means the following.

A. We favor tactics over tools, and strategies over tactics.

B. Yet we also favor investigation over rigor. In other words, we want to understand why things are true, rather than how.

C. "How" arguments are rigorous and have clear details, but the global picture is murky.

D. In contrast, a "why" argument is:

1. Not always rigorous.

2. Sometimes not even correct!

3. Globally clear, even if missing some details.

4. Magical yet inevitable.

5. Often a surprising yet natural point of view.

IV. Here are some examples from our course and elsewhere.

A. The proof that 8 times a triangular number equals a perfect square (Lecture Six) is a typical "why" argument.

B. The bug problem (Lecture Eight) can be solved analytically, with differential equations. This is a "how" argument, in contrast to our "why" argument.

C. "Why" is at the heart of most mathematicians' Platonic beliefs.

D. Problems fall all along the spectrum from completely opaque to completely understood in the "why" sense.

E. The Shanille O'Keal problem from Lecture Nineteen was a good example of a "how."

F. An example of something without even a "how," because all we had was a false conjecture, was the 5 circles problem of Lecture Four.

V. What do the "why" arguments have in common? What can we learn from them?

A. Pictures ($8T + 1$ problem).

B. Natural point of view and symmetry (bug problem).

C. Physical intuition (arithmetic-geometric mean inequality).

D. Using a physical object (Fermat's little theorem, Shanille O'Keal).

E. Dynamic visualization of lines, evolving structures, and using important combinatorial facts (5 circles).

VI. So what's next?

A. Keep learning facts.

B. But do not forget the need to build up flexibility, visualization, recasting, physical intuition, and the ability to see a natural point of view.

C. Learn about complex numbers—they incorporate all of these ideas.

D. Problem solving is not just a textbook subject—it is a lifestyle, with a culture.

E. The true underpinnings of this culture are passion and persistence.

Suggested Reading:

Aigner and Ziegler, *Proofs from THE BOOK.*

Lansing, *Endurance.*

Needham, *Visual Complex Analysis.*

Zeitz, *The Art and Craft of Problem Solving*, sec. 4.2.

Lecture Twenty-Four—Transcript
How versus Why—The Final Frontier

Welcome to the final lecture of the course. Although this is the end of the course, I hope that it's just the beginning of your involvement with problem solving. Every semester at the University of San Francisco, I get sad on the last day. I apologize to my students because I always feel that we never learned quite enough math—that there's always lots more great math out there that we have to experience. I feel the same way now here, and I have kind of a paternal worry about my students. I feel a little bit like Yoda did in *The Empire Strikes Back* when Luke Skywalker left the planet to go fight Darth Vader. Yoda said, "He is not ready." That's how I feel, but I'm going to make the optimistic assumption that you are all fired up about problem solving and you want to do more. Thus, you probably want to know about books, Web resources, collections of problems, and things of that kind, but this can easily be handled by the written materials that come with this course. I will recommend just 2 books toward the end of this lecture, but I do want to leave you with 3 things—first, some parental reminders of the "eat your vegetables" variety. Second, I want to mention a new mathematical topic that I'd like you to begin exploring, and third, I would like to suggest to you a mission to occupy you for the rest of your life.

Here's the reminder: We've learned a lot about strategies and tactics, and I'm hoping by now that you have a pretty good sense of strategies and that they're, to some degree, internalized. But tactics are always a little more complex. They're a little more narrowly focused, and it's harder to know what to do. I just want to remind you of a few of them and of a few contingencies. The real important thing is just to think about the question of contingencies. When do you use what? For example, when do you use proof by contradiction? Whenever you have a situation where you're trying to explain something that involves entities that are really slippery, such as infinity, then if you negate them, you'll get things that are less slippery, such as finite structures. Proof by contradiction is often the way to go. In fact, proof by contradiction is usually your proof of choice. Start with that assumption and see if you can work that into a logical argument.

What about the extreme principle that you have learned? Again, ask yourself if your problem involves entities where being on the boundary of them would make the situation simpler, giving you fewer degrees of freedom, thereby giving you more information. Then, try an extremal argument. Pigeonhole principle—when do you use that? It's a little bit less common, but if you're trying to prove something or to do something where the penultimate step might be that 2 things lie in the same category, then it's a natural for the pigeonhole principle. In general, whenever a problem has 2 numbers in it and one number is bigger than the other, you might wonder if it's just a contrived pigeonhole exercise of some sort. Of the proof methods, as I said, contradiction is number 1, but induction, as we've learned, can be very important as long as you're dealing with an evolving structure where you might have some sort of recurrence going on. Look for structures, especially structures that are not numerical, where you might not notice this, but structures where something large builds upon something small. Often, an inductive argument is what you need.

Finally, invariants—when do you use invariants? The flippant and simple answer is always. Always look for invariants because among invariants is symmetry, and you always want to ask yourself the question: Could symmetry be involved in my problem? The secondary question, the larger question really, is: Could invariants be used? The real thing is ask these questions. That's more than half of the battle. That was my reminder.

Now, I want to talk to you about a new mathematical topic that we're just getting to now. It's one that I'm not going to go into in much detail, but the importance of this is extreme. Here's a parable, a science fiction story I read years ago called "Earth Abides." It was written in the 1950s, and it was about an epidemic that wipes out almost everyone on Earth. The protagonist in the story is an intellectual who's trying to start a new society with a few survivors. They have babies and grandchildren, and they live a fairly primitive life, but he has access to the world's great libraries. He's trying to think what can he give his descendants because his descendants are not learning to read. They're descending into savagery, and he wants them to have some sort of an edge, some way to survive. He discovers, in his reading, that a really great invention to teach the grandchildren about is the bow and arrow because it's an invention that's very hard to come upon randomly, but it's incredibly practical, incredibly powerful, and very, very versatile. My variant of a bow and arrow for you is the mathematical topic of complex numbers.

Why complex numbers? Think of them as yet a new playground for you, like graph theory was, but this is an algebraic, geometric, physics playground, which has an incredible potential for connecting many branches of math when properly studied. The reason is is that complex numbers are like a 4-in-one package. A complex number is a number, and you can do algebra with it and even calculus with it. It's also a location in the plane once you plot complex numbers in a plane. Therefore, complex numbers can be thought of as vector quantities. But that's not all. They also are transformations because if you multiply by the complex number i, the square root of -1, that will rotate you by 90 degrees counterclockwise. So complex numbers are transformations, vectors, locations, and numbers that you can do algebra with. There's incredible potential there. The complex numbers, you can think of them as the natural numerical system for recasting between algebra and geometry, between dynamical things and static things—incredibly important for turning mathematical problems into physics problems and vice versa. Because you have this transformational aspect, complex numbers have a rich, rich connection with symmetry. In fact, complex numbers have a rich connection with just about every branch of mathematics, including number theory.

Here's a tiny example to whet your appetite—the roots of unity. It even has a sort of highfalutin name. The roots of unity are just the roots of a polynomial equation, such as $x^5 = 1$. So $x^5 = 1$ has 5 roots. One of them is 1. The other 4 are complex. We actually found them when we solved the equation $x^4 + x^3 + x^2 + x + 1 = 0$, but in the complex plane, they're much easier to understand because they form a regular pentagon if you include the real root of 1. They also form a geometric series. The roots are equal to 1, z, z^2, z^3, and z^4 where z is just a single complex number. This example shows you that we have a connection between a polynomial, and a polygon, and the idea of a rotation, and algebra, and symmetry all connected together. In addition, if you think about roots of unity in general, you basically get De Moivre's theorem, which you may have learned in school, for free. De Moivre's theorem states that $(\cos t + i \sin t)^n = \cos(nt) + i \sin(nt)$. You almost get for free Euler's famous formula $e^{it} = \cos t + i \sin t$. For this, we need a little bit of calculus but only the tiniest bit, and this leads to perhaps the most famous equation in all of mathematics, $e^{i\pi} = -1$.

The important thing is you want to learn not just how all of this is true but why this is true. That gets us to our mission. As you know, we have constantly strived to think about problem solving at a very, very high level, always taking the bird's-eye view whenever we can. We favor tactics over tools and strategies over tactics. Yet we also favor investigation over rigor. In other words, we want to understand why things are true rather than just how, but we don't want to stop thinking about things until we're completely satisfied about the why versus the how. We want to get our hands dirty until we understand why. That's your mission—to find out why and never be satisfied if you only know how. Like many seemingly obvious strategies and tactics that we've discussed in this course, the why versus how dichotomy sounds simple, but it's not. Let's learn a little bit more about it and look at some examples that will provide some closure with what we've already done in this course.

What do I mean by "how"? You've seen it a lot, but I just want to define it a little more. A how argument is always rigorous, logically rigorous, and it is what I call locally clear. In other words, line by line, every line makes sense. There are lots of details, but often there might be sort of a "black box" happening, a *deus ex machina*–type phenomenon.

Conway's checker solution is perhaps an example of a how at least for people at a relatively low level of mathematics. For someone like Conway, it's sort of a trivial why. Again, just like problems and exercises, the boundary between how and why is a little bit porous. Let's contrast how with why. A why argument is not always rigorous. In fact, it's sometimes not even correct, but it's globally clear. It's holistically clear, and it's often unexpected. It's magical, yet it has a feeling of inevitability. Often it involves a surprising but in the end very natural point of view change, in other words, some sort of a recasting, perhaps.

Let's look at a few examples from the course that were, in my opinion, very nice examples of why. The $8t + 1$ proof, where we showed that 8 times a triangular number plus 1 is a perfect square. We did it without any mathematics really, just pictures, without any algebra, at least. The 4 bugs problem, where the bugs crashed into one another, our explanation used symmetry and a natural point of view. The traditional way this is worked out is using differential equations. It's quite a hard problem using differential equations, and there is no illumination to the method. You get an equation that gives you the position of any bug at any given time, which is nice, but you don't really know why it worked. You just put it into a differential equations machine, and it cranks out a solution.

Another why that we did was our proof of the arithmetic-geometric mean inequality. I would contrast that with an induction proof that exists, which is quite ingenious, and it's fun to look at. But as you've seen sometimes, induction often is not illuminating. Induction sometimes just ends up with algebra going from one step to another. Remember our proof of Fermat's little theorem, where we recast numbers as necklaces and we changed Fermat's little theorem into counting necklaces. This, again, is a way of explaining why this theorem is true, and one can contrast this with lots of other proofs that require more advanced knowledge. In general, the more elementary, the closer you are to what I would call why.

I'd like to quote from one of my favorite mathematics books. This is *Visual Complex Analysis* by Tristan Needham, and in the preface to the book is a statement of the why philosophy, which I completely agree with. Here's the quote:

> There is one "sin" that I have intentionally committed, and for which I shall not repent: many of the arguments are not rigorous, at least as they stand. This is a serious crime if one believes that our mathematical theories are merely elaborate mental constructs, precariously hoisted aloft. Then rigor becomes the nerve-racking balancing act that prevents the entire structure from crashing down around us. But suppose one believes, as I do [and I do, too], that our mathematical theories are attempting to capture aspects of a robust Platonic world that is not of our making. I would then contend that an initial lack of rigor is a small price to pay if it allows the reader to see into this world more directly and pleasurably than would otherwise be possible.

Of course, when he says reader, he means reader because it's a book. I mean viewer. I mean you. I agree with everything that I just said here.

Let's look at mathematics now from this how versus why border. We're at, so to speak, the final frontier, and we'll look at all the math ahead of us in terms of where it lives in a continuum on the how versus why scale. At the lowest end are the problems that are completely opaque, problems that we can, in the next stages, make educated guesses about, but we're just not sure. We still don't quite understand them. Then, there are the things we actually know, but we only know how, not why. Finally, the gold standard is the things that we know why.

Here's an example of the opaque end of the spectrum. This is a quote from Erdös about the difficulty of computing Ramsey numbers. Recall how hard it was to understand things like Ramsey 3, 3, 3, the 17-gon problem, and I alluded to the fact that the numbers get more difficult. This is what Erdös had to say. He tended to have lots of very funny, quotable stories.

> Suppose aliens invade the Earth and threaten to obliterate it in a year's time unless human beings can find the Ramsey number for red 5 and blue 5. We could marshal the world's best minds and fastest computers, and within a year we could probably calculate the value. If the aliens demanded the Ramsey number for red 6 and blue 6, however, we would have no choice but to launch a preemptive attack.

This quote was some years ago, but since that time, there has been no progress on these problems and no aliens to spur us on either.

Here's an example of the educated guess variety. A number theory conjecture—it's not a theorem yet—called Goldbach's conjecture, which was first made in 1742, states the very innocent conjecture that every even number greater than or equal to 4 can be written as a sum of 2 primes. You can get your hands dirty, and many of the world's mathematicians have done just that, and go into the trillions and never find an exception to this conjecture. There is no proof as of today, but there are a lot of preliminary theorems that suggest that this is probably true. Virtually any number theorist today would be willing to bet you $10 that Goldbach's conjecture is actually a theorem that will be proven some day, but it hasn't been. A lot of mathematics is at this stage. A lot of it is at the opaque stage.

Then, of course, there is the how but not why stage of mathematics, and most of current mathematical research that you read in papers is how. There are lots of theorems that are proven out there where it's almost impossible to understand why they're true. What you're really doing, as Tristan Needham said, is you're holding this precarious structure aloft, trying to keep things going rigorously but not understanding the big picture. Most badly written textbook math is of the how variety.

What we should do is look at some whys. Why don't we go back to some things that we thought about earlier in our course, which were at the how stage, and let's try to turn them into whys. For example, remember the Shanille O'Keal problem, where we were looking at a basketball player who gets her first basket and then misses her second basket. Thereafter, the probability that she gets a basket is equal to the proportion of baskets she's already gotten. After the first basket, she misses, so on the third toss, she has a 50/50 chance of getting a basket because she has gotten 1 out of 2 baskets so far. Then suppose she gets a toss. Then she will have made 2 baskets. On the fourth toss, she'll have a 2 out of 3 chance of making a basket and so on. That's the setup for the problem. It turned out when we did experimentation, that at any stage after, say, 50 tosses, she could have anywhere from 1 to 49 baskets and they all had equal probability. We proved this with induction using algebra, and it was locally completely true. Every bit of our argument was correct, but there was no illumination. There was no understanding why it was true.

Whenever I'm confronted with a problem where the answer is something like all the probabilities are the same or all the values are the same, it tells me there has to be something fundamental going on. It's important to try to find a why for this. I struggled pretty hard, but I came up with what I hope is a way to at least explain a little bit about why Shanille O'Keal has this probability distribution that she does. Let's convert the problem. Instead of baskets, let's use bricks. Think of the process of making baskets as whenever a basket is made, we're laying a brick and we're coloring it. Think of this rectangle here. Its width is probability 1. The very first colored strip is her first basket. The second uncolored strip is her first miss. Then, we divide toss 3, the next layer, exactly in half: With 50% probability, which is colored blue, she'll get a basket, and then with 50% probability, which is colored white, she won't get a basket. Then we continue. Start with the blue on the left of that third toss. Below that, there's now a 2/3 chance she's going to make a toss. Below that, the bricks are 2, 2, 1 in size, but under the white in level 3, the bricks are 1 to 2 in size because there's only a 1 in 3 chance she's going to make a shot. We keep on drawing this picture.

If you now go down vertically through this rectangle and just count the layers of blue that you see, that will be the total score. If you go to the very left, you can see in my picture, that's going to be a score of 4 points. Then, the very next chunk is going to be a score of 3 points. Notice because the bricks are stacked in such a way that you always have exactly 3 layers of blue. Finally, you'll have 2, and then at the very end, 1. You can see that these widths are equal. The question is: Why do they end up being equal? We can do the induction, in a sense. Why don't we try to go from toss 7 to toss 8? If you look at toss 7, we divide this into zones where the score is 6, and then 5, and then 4. I haven't colored in the toss 7. Those are equal areas of 6 points, 5 points, and 4 points. I'm leaving out the rest of the diagram. Below the 6 point, we're going to have a blue to white in the ratio of 6 to 1. Below the 5, it will be in the ratio of 5 to 2, and below the 4, it will be in the ratio of 4 to 3. If you just think of those blue zones as units, you'll have 6 units and then 1 white, and then 5 units and 2 white, and then 4 units and then 3 white. If you then stagger it, you'll see that you first have 6 blues altogether. Then you have 6 units altogether of which 5 are blue. Then you have 6 units altogether of which 4 are blue, and these are all exactly the same width. They're exactly enough to just add 1 more to each of the total point values. We're going to have exactly the same amount of 7s, as 6s, as 5s in the sum, and that's how the induction works. It works by drawing pictures.

Let's go back to another problem where we didn't quite have a how, but we had an uneducated guess. Let's turn this into a why. This was the problem of the 5 circles, where we started to conjecture that the number of areas was powers of 2, but it went 1, 2, 4, 8, 16, and then 31. This was a very sad discovery we made back in Lecture Four, and the question is: What is going on, and why do we see what we see? Let's try to explain it, and the way we'll explain it is by giving some life to the geometrical entities. Let's imagine that lines can talk. What we'll do is take the case with 5 points and watch it evolve slowly but surely, but we'll watch it evolve from the point of view of lines. I start by having a blank circle, which has one area. Then I add one line. Each time I add a line, a new line, it's red. So I had one area with one line. Then I add another line, getting another area, and then I add another line, getting another area. I keep on just getting areas per lines, but eventually, we have something different happen. If you look at the third picture in the top row, that's the first time that a line intersects one of the old lines. When that happened, we got an extra area. Then, the next one, the new line intersects 2 lines, getting if you look at it, 2 more new areas rather than [one] before, a total of 3 areas. Normally, a line just adds one more area, but now we're getting more areas. What's going on?

If you remember from our induction lecture where we were watching the number of areas that lines divide the plane into, we understand this phenomenon actually, and we can use it here. A way to make it really sing to you is to imagine that as the line enters the circle and then when it exits, it sighs because it's happy to leave. Whenever a line sighs, we get a new area, and then whenever a line gets punctured by another line, it yells because it's hurt. Whenever you hear a yell, there's a new area. That's your formula. Your formula is going to be the number of sighs plus the number of yells plus 1 because the starting point when you had nothing was a circle with one area. Thus, the formula is equal to 1, the circle to start with, plus the number of lines, plus the number of interior intersection points. Now it's just a matter of computing those values. How many lines will there be? Well, remember we started with 5 points, and so for any 2 points, there's going to be a line. The combinatorial answer to this is the number of ways we can pick 2 points from a set of 5, which is the combinatorial number $\binom{5}{2}$, which is 10. We can find that in Pascal's triangle. How about the number of interior points?

Well, that's a little harder to figure out, but if we look at an interior point in there, the point x, if you think about it, it is completely and utterly determined by the 4 vertices, a, b, c, and d, of the quadrilateral whose diagonals intersect to form x. We take x, look at its parents, those diagonals, and look at their parents, the 4 vertices. Every set of 4 vertices will determine exactly one interior point. Again, it's a combinatorial question of how many ways can we choose 4 points out of our 5, $\binom{5}{4}$, again a Pascal triangle number. In general, the formula for any number of points is $1 + \binom{n}{2} + \binom{n}{4}$, but now the question is: Why did we get those powers of 2, and why did they stop? Well, here's a simple explanation. Let's try $n = 6$. Just use Pascal's triangle as your guide. If you look at the bottom row here, we see that is row 6 of Pascal's triangle, and the red numbers are 1, and then $\binom{6}{2}$, and then $\binom{6}{4}$. Remember how Pascal's triangle is defined. It's defined by summation, and if you look at the row above it, that 1 can just slide up. The 15 is equal to 5 + 10. The other 15 is equal to $10 + 5$, and what we see is that that innocent sum of 3 numbers is always going to equal the sum of the first 5 elements in a row of Pascal's triangle. If Pascal's triangle has more than 5 elements, we won't quite get to the sum of the rows, but we are old hands with Pascal's triangle. We know that the sum of the rows is 2^n. It sort of mimicked 2^n for a while, but what we're really seeing is not the 2^n but just the sum of 5 terms in Pascal's triangle. That's why this formula is what it is. It's why we see what we see.

What are the big lessons we can learn from this how versus why dichotomy? Let's go back and take a very quick look at those examples. The $8t + 1$ example—pictures. The bugs example—symmetry, point of view, and even thinking like a bug. The arithmetic-geometric mean—physical intuition. Fermat's little theorem—again, using a physical object to solve a problem. The Shanille O'Keal problem—physical stacking of bricks to visualize probabilities. The 5 circles problem—dynamic visualization, not seeing lines in a static way but seeing them move and then seeing them talk. Knowing, also, facts—knowing about combinatorics, knowing about Pascal's triangle.

What should we work on now? What does this tell us? What's the program for us to do in the future? Well, again, your mission is why, so to build up your why program this is what you should focus on: imaginative recasting, especially using physical intuition, especially looking for dynamic solutions rather than static solutions, which makes you think about physics. Even if you don't know much about physics, remember the marble problem with angular momentum. You don't need to be an expert physicist to understand that problem. Always work on visualization. Try to think about pictures, but also think about point of view changes, and then start learning about complex numbers because they incorporate all of these ideas. Remember, we learned that problem solving is not just a textbook subject, but it's a lifestyle with a culture. You have been admitted to what I call the nerd culture, but I don't expect you to run out and memorize the first 100 digits of pi. If you want to, you can, but I don't expect you to. It's not necessary because the true underpinnings of this culture are passion and persistence.

Let's do a little cross-training. I want to recommend an inspirational book, which will help solidify your position in this mathematical culture. The book is not a math book. It's *Endurance* by Alfred Lansing, a tale of the Shackleton voyage, which was shipwrecked in Antarctica for several years. What it is is a tale of mental discipline, creativity, and calmness. It's a wonderful inspiration for mathematical problem solving. Now, for mathematical inspiration, I hope that you will begin your study of complex numbers with *Visual Complex Analysis*, the book by Tristan Needham whose preface I read to you. This is the only book that really explains, in my opinion, the why of complex numbers, and it's also an ideal book for somebody who is perhaps a mathematical amateur to read because the first few chapters don't use any calculus. They involve geometry and algebra and always hunt for the simplest, most elementary explanations possible.

Now what? I've given lots of mountain-climbing analogies, so I want to end with one more climbing analogy. When I was first learning rock climbing when I was a high school teacher in Colorado, I was really nervous about preparation. I asked my teacher, "What do I need to do to get good at this? Should I work on doing pull-ups with my fingers, that sort of thing?" The teacher laughed and he said, "No, no, no. There are only 2 things that you need to do, and you need to do these every day. Every day, practice your knots, and every day, do stretching exercises." I want to remind you to do the same thing. Keep practicing math. Learn facts like Pascal's triangle, but stay flexible and open to new points of view. I just want to end with one thing. Remember, Yoda was worried about Luke Skywalker not being ready, but in the end, he was ready and he lived happily ever after.

Solutions

Lecture Thirteen

1. By Fermat's little theorem, $3^{18} = 1 \pmod{19}$. Since $18 \times 111 = 1998$, by raising the previous equation to the 111th power, we get $3^{1998} = 1 \pmod{19}$. Notice that $3^{2009} = 3^{1998} \times 3^{11}$. Hence our answer will be whatever 3^{11} equals modulo 19. We compute $3^2 = 9$, $3^3 = 27 = 8$, $3^4 = 24 = 5$, and $3^5 = 15 = -4$. Squaring this last one, we get $3^{10} = 16 = -3$, and then finally $3^{11} = -3 \times 3 = -9 = 10$ (all mod 19).
2. If it was a line instead of a necklace, there would be 6 choices for the first bead, 5 for the second, and so on, for $6 \times 5 \times 4 \times 3 \times 2 \times 1 = 720$ choices. But since it is a necklace, we can take any arrangement such as *bcadfe* and cyclically permute it, say, to *cadfeb*. There are 6 members in each sorority, so we divide by 6, getting 120 as our answer.

Lecture Fourteen

1. Partition the unit square into four $\frac{1}{2} \times \frac{1}{2}$ squares. By pigeonhole, one of these smaller squares must contain at least 2 points. Since the diagonal of each small square is $\frac{\sqrt{2}}{2}$, that is the maximum distance between the 2 points.
2. There are 150,001 categories (include bald people!). We need to find the minimum population p such that the ceiling of p divided by 150,001 is equal to 10. Thus $p = 9 \times 150{,}001 + 1 = 1{,}350{,}010$.

Lecture Fifteen

1. Getting your hands dirty should yield the formula $v - e + r = 1$. This is called Euler's formula.
2. You cannot. If you could, then the quotients $(b - c)/(a - b)$, $(c - a)/(b - c)$, and $(a - b)/(c - a)$ would all be integers. However, the product of these 3 quotients is invariant; it is 1. The only way that 3 integers can multiply to 1 is if all of them are equal to 1. But that would mean that $2b = a + c$, $2c = a + b$, and $2a = b + c$. In other words, a, b, and c are each the average of the other 2. Using the extreme principle, let, say, a be the smallest of the 3 numbers (remember, the numbers are distinct). Then we have a contradiction. How can a be the average of 2 numbers that are both larger than a? The other possibility is if one of the quotients is 1 and the other 2 are −1. But if a quotient equaled −1, for example, the first quotient, we would get $b - c = b - a$, which makes $c = a$, another contradiction.

Lecture Sixteen

1. Reflect across the stream. Then when you build your optimal rectangle with perimeter S, the mirror rectangle also has perimeter S, and collectively, you are building a rectangle with 4 sides that is to have maximal area, with a fixed perimeter of $2S$. You know that this optimal shape is a square. So the answer to the original question is a half square: a rectangle whose sides are in the ratio 1:2, with the longer side on the stream.
2. Just multiply the 3 inequalities $(a+b) \geq 2\sqrt{ab}$, $(b+c) \geq 2\sqrt{bc}$, and $(c+a) \geq 2\sqrt{ca}$, and you are done.

Lecture Seventeen

1. Imagine, on the second day, a clone of the first monk who starts at the bottom at 8 am and exactly imitates what the first monk did the day before. Clearly the 2 monks will meet on the trail, and this time and place are the solution.
2. Draw a lattice of squares. Start at (0, 0), and begin drawing a line with slope 7/11. As it goes northeast, it will first hit another lattice point at (11, 7). To count bounces, just count the number of times this line hits a horizontal or vertical lattice line. It will hit $x = 1, 2, 3, \ldots, 10$ and $y = 1, 2, 3, \ldots, 6$, for a total of 16 bounces.

Lecture Eighteen

1. Just rotate the diagram 60° clockwise about the center K. Then J moves to M, and N moves to I. Hence $JN = IM$. Likewise, a rotation about I will show that $LK = JN$.
2. Start with the point (0, 0). The first rotation leaves it fixed; the second one, about (1, 0), brings it to (1, −1); the third brings it to (3, −1); and the final one brings it to (4, 0). So the translation (of any starting point) will be "move 4 units to the east." You might conjecture that if you have n rotations, about (0, 0), (1, 0), … , $(n - 1, 0)$, each by $\frac{360}{n}^{\circ}$ counterclockwise, the net result would be a translation by n units to the right, and this is correct. The easiest way to see it is to imagine a regular n-gon with side length 1 unit whose "top" side is the line segment joining (0, −1) and (0, 0). Then the rotations are equivalent to rolling this n-gon along the x-axis, ending up n units to the right. Try this for $n = 3$, $n = 4$, or $n = 5$ to see for sure.

Lecture Nineteen

1. It is easy to conjecture that the sum is $F_{n+2} - 1$. The base case is clearly true ($1 = F_3 - 1$). For the inductive step, assume that it is true that the sum of the first k Fibonacci numbers is equal to the $(k + 2)^{\text{th}}$ Fibonacci minus 1. Then we want to find the sum of the first $(k + 1)$ Fibonaccis. This sum is $F_1 + F_2 + \cdots + F_k + F_{k+1}$, which by the inductive hypothesis can be written as $F_{k+2} - 1 + F_{k+1}$. By the definition of Fibonacci numbers, this is equal to $F_{k+3} - 1$, which is what we wanted.
2. Once again, the base case is obvious, since the first Fibonacci number is less than 2. Thereafter, it is simple as well: Since each successive Fibonacci is equal to the sum of the 2 preceding it, and since the sequence is an increasing one, each Fibonacci number is strictly less than twice the one before it.

Lecture Twenty

1. Just plug in $x = 10$ into the expression $(1+x)^k$. For example, if $k = 4$, we get $11^4 = 10^4 + 4 \cdot 10^3 + 6 \cdot 10^2 + 4 \cdot 10^1 + 1$. Since we use the base-10 system, that is the number 14,641. The reason it fails for greater values of k is because some of the coefficients of the binomial are greater than 9. For example, if $k = 5$ we get $11^5 = 10^5 + 5 \cdot 10^4 + 10 \cdot 10^3 + 10 \cdot 10^2 + 5 \cdot 10^1 + 1$. We cannot just read this off as a base-10 number as we did the last time, because this number has a thousands place digit of 10 and a hundreds place digit of 10. In base 10, digits cannot be greater than 9.
2. The analysis is similar to what we did in the lecture, but this time, recursive structures are successive powers of 3, and each new triangle consists of 6 triangles of the previous kind, with an inverted 0 in the middle. You can search the Web for "Pascal's triangle modulo 3" to find good illustrations and interactive applets; one example is at http://faculty.salisbury.edu/~kmshannon/pascal/article/twist.htm. The reason for the number 6 is that it is equal to 1 + 2 + 3. The limit of the number of nonzeros (mod 3) goes to 0 as the number of rows increases, so just as before, almost all elements of Pascal's triangle are multiples of 3!

Lecture Twenty-One

1. Using the multiplication rule for generating functions, we get $\binom{17}{13}=\binom{10}{6}\binom{7}{7}+\binom{10}{7}\binom{7}{6}+\binom{10}{8}\binom{7}{5}+\binom{10}{9}\binom{7}{4}+\binom{10}{10}\binom{7}{3}$. This has a good combinatorial interpretation: The left-hand side is the number of ways to pick 13 children from a pool of 17. The right-hand side counts the same thing but supposes that the children consist of 10 girls and 7 boys, and it breaks the outcomes into the 5 cases of 6 girls and 7 boys, 7 girls and 6 boys, 8 girls and 5 boys, 9 girls and 4 boys, and 10 girls and 3 boys.

2. The series is $1 + x + 2x^2 + 4x^3 + 8x^4 + 16x^5 + 32x^6 + 63x^7 + 125x^8 + \cdots$, and the coefficient of x^n is equal to the number of ways that any number of dice can add up to n. For example, there are 2 ways to get 2: two 1s or one 2. And there are 16 ways to get 5: 5; 1, 4; 4, 1; 1, 1, 3; 1, 3, 1; 3, 1, 1; 3, 2; 2, 3; 1, 2, 2; 2, 1, 2; 2, 2, 1; 1, 1, 1, 2; 1, 1, 2, 1; 1, 2, 1, 1; 2, 1, 1, 1; or 1, 1, 1, 1, 1. The reason this works is because for each k, $[D(x)]^k$ is the generating function for dice sums when k dice are rolled. You may enjoy looking up this sequence in The Online Encyclopedia of Integer Sequences; they are called the "hexanacci" numbers. Can you see why? Because they satisfy the recurrence formula that each number in the sequence is equal to the sum of the 6 previous numbers!

Lecture Twenty-Two

1. Use proof by contradiction. Assume, to the contrary, that for neither of the 2 colors is it true that one can find 2 points of the same color at any mutual distance. In other words, there are distances x and y such that there are no 2 red points x units apart and no 2 blue points y units apart. Without loss of generality, suppose that x is greater than or equal to y (i.e., use the extreme principle on 2 numbers). Now, consider a red point. There must be at least 1 red point, for otherwise the color blue would not have a forbidden distance y. Draw a circle with radius x and center at this red point. Every point on this circle must be blue, since no 2 red points are x units apart. But now we have achieved a contradiction: Since y is less than or equal to x, it is certainly possible to find 2 (blue) points on this circle that are y units apart.
2. Start by finding 2 points that are the same color, say, red, at locations *A* and *B*. Now consider the configuration below made of equilateral triangles. Besides the small equilateral triangles, note that there are larger equilateral triangles, such as *CDE* and *AEH*. The existence of these alternatives is key for a miniature Gallai-style argument: Either *C* or *D* are red; then we would be done. Otherwise, both *C* and *D* are blue. Then if *E* is blue, we are done again (triangle *CDE*). But if *E* is red, then we will be done if either *F* or *G* is red. If not, they are both blue. But now we have a focal point at *H*. If it is red, we have a large red equilateral triangle (*AEH*). If it is blue, we have the small equilateral triangle *HCF*. No matter what, a monochrome equilateral triangle is guaranteed.

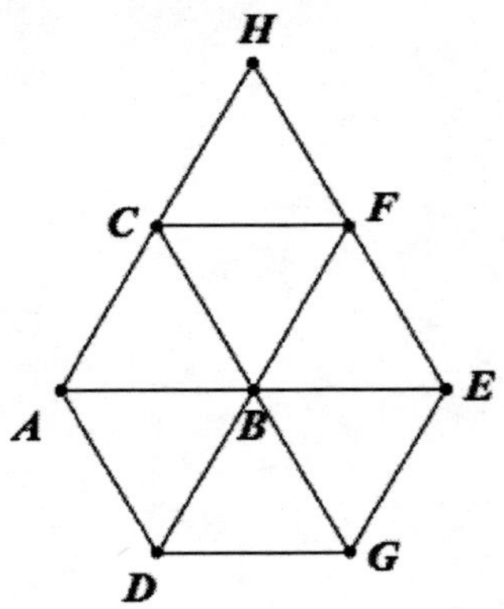

Lecture Twenty-Three

1. If we assign a value greater than 1 to the point *C*, the Conway sum is no longer a monovariant, if you look at moves involving *C* itself. For example, if a checker 2 units to the left of *C* jumped over a checker 1 unit to the left of *C* and then occupied *C*, then the Conway sum would actually increase!
2. Model it with a graph: Use vertices for mathematicians and edges to indicate joint papers. There is only one graph that satisfies the conditions.

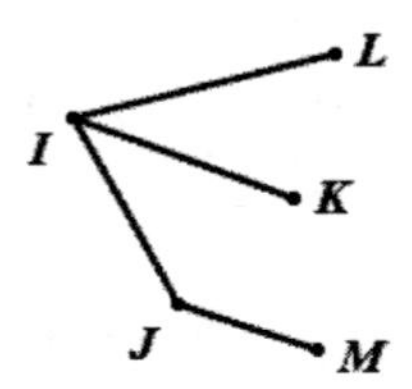

The Erdös numbers depend on which one is Erdös. If Erdös is person *I*, then *L*, *K*, and *J* have Erdös number 1, and *M* has Erdös number 2. But if we put Erdös anywhere else, it is possible for someone to have an Erdös number of 3.

Lecture Twenty-Four

Not applicable.

Timeline

1787 .... Ten-year-old Carl Gauss uses Gaussian pairing.

1796 .... Teenaged Carl Gauss solves the problem of how to construct the regular heptadecagon (17-gon), something that had eluded mathematicians since Euclid's time.

1832 .... French algebraist Évariste Galois dies in Paris from a duel at the age of 20.

1894 .... The Eotvos contest, the first Olympiad-style math competition, begins in Hungary.

1907 .... Willem Abraham Wythoff writes about the game later called Wythoff's Nim, actually an older game that probably originated in China.

1930 .... Frank Ramsey's paper "On a Problem in Formal Logic" is published; Tibor Gallai's theorem is proven around the same time.

1938 .... Annual Putnam competitions for college undergraduates begin.

1939 .... Physicist Richard Feynman is named a Putnam Fellow.

1952 .... High school mathematics contests sponsored by the Mathematical Association of America begin across the U.S.

1959 .... The first International Mathematical Olympiad is held in Romania, with teams from 7 countries.

1961 .... John Conway analyzes the checker problem.

1970 .... John Conway invents the cellular automaton Game of Life, popularized by Martin Gardner in *Scientific American*.

1972 .... The first USA Mathematical Olympiad is held.

1974 .... The U.S. participates for the first time in the International Mathematical Olympiad (in East Germany), placing second after the USSR.

1977 .... Wythoff's Nim is popularized by Martin Gardner in *Scientific American*.

1981 .... The International Mathematical Olympiad is held in the U.S. for the first time, with 27 countries participating.

1985 .... The Colorado Mathematical Olympiad begins, founded by Soviet émigré Alexander Soifer. This was the first regional Olympiad-style contest in the U.S.

1994 .... The U.S. team comes in first place at the 35th International Mathematical Olympiad (held in Hong Kong), with all 6 team members receiving perfect scores. This was the first and only time a team has received a perfect score at that competition.

1996 .... Hungarian-born Paul Erdös, the most prolific mathematician and problem solver of modern times, dies.

2001 .... The U.S. hosts the International Mathematical Olympiad for the second time, with 83 countries attending.

Glossary

algorithmic proof: Proof where we imagine a sequence of steps that is guaranteed to solve our problem.

binomial coefficients: The numbers $\binom{n}{k}$, which are equal to (1) the coefficient of x^k in $(1+x)^n$, (2) the number of ways of choosing k things from a set of n things, and (3) the number $n(n-1)\cdots(n-k+1)/k!$.

bipartite graph: A graph whose vertices can be colored red and blue in such a way that no edge connects vertices of the same color.

congruence: Two integers are said to be congruent (modulo m) if their difference is multiple of m.

crux move: The crucial step in a problem-solving investigation that solves the problem. This step can be technical or can be a strategic breakthrough.

degree: A graph theory term; the degree of a vertex is the number of edges emanating from it.

extreme principle: The problem-solving tactic that says, "contemplate the extremal values of your problem."

Fibonacci numbers: The sequence 1, 1, 2, 3, 5, … in which each term is equal to the sum of the 2 previous. Named after Leonardo Fibonacci (c. 1170–c.1250), it is one of the most accessible playgrounds of recreational mathematics.

fundamental theorem of arithmetic: All integers can be factored into primes, and this factorization is unique (up to order).

generating function: Given a sequence $a_0, a_1, a_2, \ldots$, its generating function is the polynomial $a_0 + a_1x + a_2x^2 + \cdots$. The generating function encodes information about the entire sequence; algebraic manipulations of generating functions can thus shed light on questions about the sequence.

Goldbach conjecture: A famous unsolved problem asserting that all even numbers greater than 2 can be written as a sum of 2 primes.

golden ratio: The number $(1+\sqrt{5})/2$, which crops up in many places in mathematics, including the Fibonacci numbers.

graph theory: The branch of math that studies abstract networks, also known as graphs, which are entities of vertices joined by edges. It is easy to learn and therefore a very accessible laboratory for exploring a number of problem-solving themes.

handshake lemma: An important graph theory result stating that the sum of the degree of each of the vertices of a graph is equal to twice the number of edges in the graph.

harmonic series: The sum of the reciprocals of the positive integers, which is a divergent series (meaning the sum is infinite).

induction: Technically called mathematical induction, this is a powerful method of proving recursive statements. An inductive proof always has 2 parts: A base case proving the first stage is followed by the inductive step, where it is shown that each intermediate stage logically implies the next.

integers: The positive and negative whole numbers, including zero.

International Mathematical Olympiad (IMO): An elite problem-solving contest, initiated by Eastern Bloc countries in 1959, that now includes nearly 100 nations every year.

invariant: A very high-level tactic for looking at many problems where a quantity or quality stays unchanged. A monovariant is a quantity that changes in only one direction; monovariants are very useful for studying evolving systems and proving that they terminate or that certain states are impossible.

modulus: A mathematical entity that in a congruence divides the difference of 2 congruent members without leaving a remainder. *See also* **congruence**.

monovariant: *See* **invariant**.

number theory: The branch of math dealing with integers. Because integers are familiar to us beginning in grade school, number theory, like graph theory, is an excellent venue for mathematical investigation for beginners.

Olympiad: A style of math contest that features relatively few questions (usually fewer than 10) of the essay-proof type.

parity: The property of oddness or evenness for an integer. Parity is a powerful tactic that reduces a problem from a large, or infinite, number of states to only 2 states.

Pascal's triangle: Perhaps the greatest of all the elementary mathematical laboratories for investigation. This is a triangle of numbers where row n consists of the binomial coefficients $\binom{n}{k}, k = 0,1,2,\ldots,n$.

pigeonhole principle: A fundamental problem-solving tactic stating that if you have more things (pigeons) than categories (pigeonholes), at least 2 of the things belong to the same category.

prime number: A positive integer that has no positive integer divisors other than 1 and itself. The first few primes are 2, 3, 5, 7, 11, and 13.

problem: A mathematical question that one does not know, at least initially, how to approach and that therefore requires investigation, often using organized strategies and tactics. We contrast this with an exercise: a question that may be difficult but is immediately approachable with little or no investigation.

proof by contradiction: A method of proof that starts by assuming that the conclusion is false and then proceeds to a logical contradiction, concluding that the conclusion's falsehood was untenable.

Putnam exam: An Olympiad-style contest for American and Canadian undergraduates, notorious for having a median score, most years, of 0 or 1 out of 120 possible points.

Pythagorean theorem: The sum of the squares of the legs of a right triangle is equal to the square of its hypotenuse. There are hundreds of known proofs of this theorem, the earliest thousands of years old.

Ramsey theory: A branch of mathematics named in honor of Frank Ramsey (1903–1930), whose seminal 1930 paper "On a Problem of Formal Logic" began the subject with a statement and proof of what is now called Ramsey's theorem. The theorem essentially asks the question, how large must a structure be in order that it is guaranteed to contain a specified substructure? The pigeonhole principle is the trivial case of Ramsey's theorem, and Gallai's theorem about squares is an example of a Ramsey-like theorem.

recasting: The problem-solving strategy of radically changing the venue of a problem, for example, from number theory into geometry, or vice versa. Certain mathematical ideas, such as generating functions, are useful precisely because of their recasting potential.

recursive definition: A sequence or evolving structure where the later terms (or more complex structures) depend on the previous, simpler ones. The Fibonacci numbers are a simple example; the chessboard tromino problem of Lecture Nineteen is another.

strategies: Mostly commonsense organizational ideas that help overcome creative blocks to begin and facilitate a problem-solving investigation. Strategies in this course include wishful thinking, make it easier, get hands dirty, chainsaw the giraffe, draw a picture, change your point of view, and recast your problem.

symmetry: An object (not necessarily geometric) is symmetrical if a transformation leaves it invariant. A natural point of view is often a point of symmetry. Symmetry increases order in a problem, so you should seek, and even impose, symmetry where possible.

tactics: Narrower than strategies, tactics are broad ideas within mathematics generally used at a later stage of investigation, often providing the key to a solution. Examples in this course include symmetry, parity, the extreme principle (contemplate extreme values), the pigeonhole principle, and squarer is better.

tools: Mathematical ideas of very narrow application that are nonetheless capable of very impressive results when used correctly. Examples in this course include Gaussian pairing, telescoping, and massage. Some useful tools (e.g., creative algebraic substitution) are better understood as a narrow instance of a broader strategy (e.g., wishful thinking).

tromino: A shape made out of 3 contiguous square units. There are only 2 types of tromino, the L and the I (a straight line of 3 squares). Trominos and more complex shapes (e.g., the 12 different pentominos, made of 5 squares) are popular objects of study in recreational mathematics.

United States of America Mathematical Olympiad (USAMO): The first national Olympiad of the United States, which began in 1972 with funding from numerous mathematical societies and the Department of Defense. It is the culminating exam that begins with multiple-choice tests taken by several hundred thousand high school and middle school students. Since 1974, the top scorers on this exam have competed for places on the 6-person team to the International Mathematical Olympiad.

Wythoff's Nim: One of the many names of a simple combinatorial game whose solution involves Fibonacci numbers and the golden ratio. Nim is an ancient game in which 2 players typically take turns removing objects from piles until none are left.

Biographical Notes

Conway, John Horton (b. 1937): A British-born mathematician and professor at Princeton University who is famous for fundamental contributions to many branches of mathematics, including recreational math.

Erdös, Paul (1913–1996): The most prolific mathematician of modern times (perhaps ever), Erdös was also known for collaborating with more mathematicians than any other. He was famous for his deliberately homeless and celibate life, devoted entirely to mathematics.

Euler, Leonhard (1707–1783): A Swiss mathematician, about as prolific as Erdös for his time, who was the father of graph theory. Euler was known for unconventional, "rule-breaking" approaches to hard mathematical problems.

Fermat, Pierre de (1601–1665): A French mathematician who was one of the first investigators of what is now modern number theory.

Gallai, Tibor (1912–1992): A Hungarian mathematician who was an important collaborator and lifelong friend of Erdös.

Galois, Évariste (1811–1832): This French mathematician, famous for his short but productive life, died in a duel. He made seminal and highly original contributions to the algebra of polynomials, among other things.

Gardner, Martin (b. 1914): Editor of the Mathematical Games column of *Scientific American* from 1956 to 1981, Gardner is unquestionably the greatest modern expositor of mathematics writing in English.

Gauss, Carl Friedrich (1777–1855): A German mathematician who is universally recognized as one of the greatest 2 or 3 mathematicians in history. He made fundamental advances in all branches of mathematics, usually generations ahead of his time.

Klein, Felix (1849–1925): A German mathematician and influential expositor. He proposed the important point of view change that geometry is best understood by looking at transformations rather than objects.

Bibliography

The list of books and resources below is pretty large, even though it just scratches the surface of the literature out there. If you are just getting started and really want to become a better problem solver, you must get practice by working on problems, and it is best for the problems to be fairly gentle and nontechnical.

The best place to start is by perusing *any* book by Martin Gardner. The CD collection of his books is an economical way to go (and its searchable). If you feel overwhelmed by the sheer quantity of Gardner's work, there are 4 great single-volume choices. George Polya's classic *How to Solve It* is short and very useful. My book, *The Art and Craft of Problem Solving*, has a larger collection of problems and a more explicit treatment of strategy and tactic. The *Mathematical Circles (Russian Experience)* book lead-authored by Dmitri Fomen has a wealth of "easy" problems (i.e., intended for Russian middle school–aged kids) along with good pedagogical ideas. And perhaps the most enjoyable single-volume book to look at is Ravi Vakil's *Mathematical Mosaic*. It is not comprehensive, but the mathematical topics are chosen with great taste. Its style may seem a little juvenile—it was written for a young audience—but the mathematics is actually quite deep.

Aigner, M., and G. Ziegler. *Proofs from THE BOOK*. Berlin: Springer, 2000. A collection of proofs that satisfy Paul Erdös's criteria of elegance, simplicity, and beauty.

Andreescu, Titu, and Svetoslav Savchev. *Mathematical Miniatures*. Washington, DC: Mathematical Association of America, 2003. The dichotomy between exercises and problems was first made clear to me by Titu Andreescu, who was the head coach of the USA team at the International Mathematical Olympiad for many years. Titu has written and coauthored numerous books about problem solving at the highest levels; this is my personal favorite.

Beck, M., and S. Robbins. *Computing the Continuous Discretely*. New York: Springer, 2007. This is an elegant (albeit advanced) book that explores the relationship between combinatorics and geometry, mostly via counting lattice points.

Bell, E. T. *Men of Mathematics.* New York: Touchstone, 1986. First published in the 1930s, this is a classic history of mathematicians. It may not be the most accurate, but it is certainly responsible for many mathematicians' worship of figures such as Carl Gauss and Évariste Galois.

Berlekamp, E. R., J. H. Conway, and R. K. Guy. *Winning Ways for Your Mathematical Plays*. Vols. 1–2. London: Academic Press, 1982. A classic and groundbreaking exposition of the theory of mathematical games.

Conway, John H., Heidi Burgiel, and Chaim Goodman-Strauss. *The Symmetry of Things*. Wellesley, MA: A. K. Peters, 2008. The latest of Conway's many books, this one is bound to be a classic.

Edwards, A. W. F. *Pascal's Arithmetical Triangle: The Story of a Mathematical Idea.* Baltimore, MD: Johns Hopkins, 2002. A carefully researched history of Pascal's triangle, perhaps the most accessible mathematical playground for problem solvers.

Engel, Arthur. *Problem-Solving Strategies*. New York: Springer, 1998. An indispensable guide to strategic problem solving at the advanced level.

Fomin, Dmitri Sergey Genkin, and Ilia Itenberg. *Mathematical Circles (Russian Experience*). Translated by Mark Saul. Providence, RI: American Mathematical Society, 1996. This is an inspiring and eye-opening guide to what Russian 12-year-olds learn in a math circle.

Gardiner, Anthony. *Discovering Mathematics: The Art of Investigation.* New York: Oxford University Press, 1986. A beautifully written elementary guide for beginners.

Gardner, Martin. *Aha! A Two Volume Collection.* Washington, DC: Mathematical Association of America, 2006. Originally published separately and now reissued as a single volume, this is a collection of some of the short puzzles of Martin Gardner. The theme of the aha puzzles is the unexpected, creative solution.

———. *Martin Gardner's Mathematical Games.* Washington, DC: Mathematical Association of America, 2005. CD-ROM. Gardner, who edited the Mathematical Games column of *Scientific American* magazine from 1956 to 1981, is in my opinion the greatest English-language mathematical expositor of modern times. He has published dozens of books based on his columns (all now collected on this single CD); any of these books contains a wealth of recreational problems along with many fun and deep essays, always written for the intelligent layperson.

———. *Martin Gardner's Sixth Book of Mathematical Diversions from Scientific American.* Chicago: University of Chicago Press, 1984.

———. *Penrose Tiles to Trapdoor Ciphers.* Rev. ed. Washington, DC: Mathematical Association of America, 1997. One of my favorites of Gardner's many collections, this book contains, among other things, a very nice discussion of the puppies and kittens game (Wythoff's Nim).

Goodaire, Edgar, and Michael Parmenter. *Discrete Mathematics with Graph Theory.* Upper Saddle River, NJ: Prentice-Hall, 2005. A very readable discrete math text; an excellent book for beginners to learn about combinatorics and graph theory.

Graham, Ronald, Donald Knuth, and Oren Patashnik. *Concrete Mathematics.* 2nd ed. Reading, MA: Addison-Wesley, 1994. This is an encyclopedic guide to many discrete math topics, from a real problem-solver's perspective. Full solutions to a wide variety of exercises and problems are at the back of the book. It is one of the essential books in my library.

Graham, Ronald, Bruce Rothschild, and Joel Spencer. *Ramsey Theory.* 2nd ed. Hoboken, NJ: Wiley, 1990. A fascinating (albeit quite advanced) discussion of many Ramsey theory topics, by one of Erdös's favorite collaborators, Ronald Graham.

Guy, Richard K. "The Strong Law of Small Numbers." *American Mathematical Monthly* 95 (1988): 697–712. This classic essay, published in *American Mathematical Monthly* but accessible to laypeople, is a play on the law of large numbers, a statement about how long-term empirical frequencies approach theoretical probabilities. In this essay, the focus is on surprising sequences that appear to do one thing but in fact do something entirely unexpected.

Hardy, G. H. *A Mathematician's Apology*. Cambridge: Cambridge University Press, 1992. First published in 1940, this is a poetic memoir about the beauty of mathematical thinking.

Hoffman, Paul. *The Man Who Loved Only Numbers*. New York: Hyperion, 1999. A beautifully written book about an unbelievable character, Paul Erdös, the most prolific researcher in the history of mathematics.

Honsberger, Ross. *Ingenuity in Mathematics*. Washington, DC: Mathematical Association of America, 1970. This and the next book are just 2 of the many wonderful works by Honsberger, whose specialty is clear essays that explain amazingly creative mathematics.

———. *Mathematical Gems II*. Washington, DC: Mathematical Association of America, 1976.

Kazarinoff, Nicholas. *Geometric Inequalities*. Washington, DC: Mathematical Association of America, 1975. A short and brilliant book that gives beginners a real intuition about inequalities by using a geometric, visual approach that minimizes algebra in favor of true insight.

Kendig, Keith. *Sink or Float? Thought Problems in Math and Physics*. Washington, DC: Mathematical Association of America, 2008. An elementary book to help develop your physical intuition in a mathematical context.

Lansing, Alfred. *Endurance*. New York: Carroll and Graf, 1999. First published in 1959, this is a riveting account of an epic tale of survival: the ill-fated Antarctic voyage of Ernest Shackleton. It is relevant to us because of the important story of mental toughness.

Lehoczky, Sandor, and Richard Rusczyk. *The Art of Problem Solving*. 7th ed. Vols 1–2. Alpine, CA: Art of Problem Solving, 2008. An excellent guide for beginners (as young as middle school), with complete solutions to problems from a very wide variety of topics.

Liu, Andy, ed. and trans. *Hungarian Problem Book III*. Washington, DC: Mathematical Association of America, 2001. There are several volumes in English of the famous Hungarian Problems, the earliest modern Olympiad-style contest. All 3 volumes of this series have excellent commentary about problem solving in general, but this is the best of them.

Maurer, Stephen B., and Anthony Ralston. *Discrete Algorithmic Mathematics*. 3rd ed. Wellesley, MA: A. K. Peters, 2004. A superb text, especially notable for its careful treatment of mathematical induction.

Needham, Tristan. *Visual Complex Analysis*. New York: Oxford University Press, 1999. An essential book for anyone who truly wants to understand *why* things are true. Needham's uncompromisingly visual approach is unique and powerful.

Nelson, Roger. *Proofs without Words*. Washington, DC: Mathematical Association of America, 1997. One of the first books to stress the importance of avoiding algebra whenever possible. Highly recommended.

Niven, Ivan. *Maxima and Minima without Calculus*. Washington, DC: Mathematical Association of America, 2005. Like the 2 books above, this is a highly recommended antidote to using too much higher-powered mathematics when simpler methods are better and ultimately more illuminating.

Olson, Steve. *Count Down: Six Kids Vie for Glory at the World's Toughest Math Competition*. Boston: Houghton Mifflin Harcourt, 2004. A fascinating account of the 2001 International Mathematical Olympiad, which was held in the United States.

Polya, G. *How to Solve It*. 2nd ed. Princeton, NJ: Princeton University Press, 1957. This is the classic book about problem solving. All later books, mine included, owe a tremendous debt to Polya's pioneering work and writing.

Soifer, Alexander. *The Mathematical Coloring Book*. New York: Spinger, 2009. This book is a mathematical history of Ramsey-style problems involving coloring, including Gallai's theorem. It also includes much fascinating history.

———. *Mathematics as Problem Solving*. New York: Springer, 2009. A very short but intense guide to problem solving for beginners, with an excellent choice of problems.

Solow, Daniel. *How to Read and Do Proofs*. 5th ed. Hoboken, NJ: Wiley, 2009. One of the standard college texts on this difficult topic.

Sylvester, J. J. "Question 7382." *Mathematical Questions from the Educational Times* 37 (1884): 26. This is the first recorded mention of what we refer to as the chicken nuggets problem, written by an eminent British mathematician.

Tabachnikov, Sergei, ed. *Kvant Selecta: Algebra and Analysis, I.* Providence, RI: American Mathematical Society, 1999. This book is part of a series of translations from the great Russian journal *Kvant* (Quantum), which popularized deep mathematics for a young audience.

Tanton, James. *Solve This: Math Activities for Students and Clubs.* Washington, DC: Mathematical Association of America, 2001. This is a one-of-a-kind collection of deep puzzles that have a hands-on quality. Perfect for truly doing mathematics and developing physical intuition.

Vakil, Ravi. *A Mathematical Mosaic.* Ontario: Brendan Kelly, 1996. One of my favorite books for beginners, this is an absolutely joyous excursion through many topics. Vakil is one of the best contest problem solvers in history, but only modesty and generosity come across in this remarkable little book.

Vanden Eynden, Charles. *Elementary Number Theory.* Long Grove, IL: Waveland Press, 2001. One of the clearest and simplest guides to this essential topic.

Velleman, Daniel. *How to Prove It.* New York: Cambridge University Press, 2006. This excellent book is a standard college-level text.

Weyl, H. *Symmetry.* Princeton, NJ: Princeton University Press, 1983. A classic, first published in 1952, that attempts, quite successfully, to bridge the gap between mathematics and aesthetics.

Wilf, Herbert. *generatingfunctionology.* 2nd ed. San Diego, CA: Academic Press, 1994. A beautiful (advanced) book that is as much about mathematical thinking as it is about generating functions.

Yaglom, I. M. *Geometric Transformations I.* New York: Random House, 1962. An elegant introduction to the transformational way of thinking. Excellent for beginners.

Zeitz, Paul. *The Art and Craft of Problem Solving.* 2nd ed. Hoboken, NJ: Wiley, 2007. My book parallels many of the topics in this course and has a wide variety of problems of many levels of difficulty.

Films and Online Resources:

Art of Problem Solving. http://www.artofproblemsolving.com. This is perhaps the world's preeminent online community of math enthusiasts, most of them young and interested in math contests. It has numerous resources for learning mathematics.

Csicery, George Paul. *Hard Problems*. 2008. DVD. A thought-provoking documentary about the formation of the 2006 U.S. International Mathematical Olympiad team and its adventures at the competition, which was held in Slovenia. This film initially aired on PBS stations around the United States in 2009.

———. *N is a Number*. 2007. DVD. A fascinating 1-hour documentary about the life of Paul Erdös.

The Online Encyclopedia of Integer Sequences. http://www.research.att.com/~njas/sequences/Seis.html.

This encyclopedia contains more than 100,000 sequences; it is a math nerd's paradise.

Notes